Please note that the previous printing included a CD-ROM.

The material is now only available on the companion website:
http://booksite.elsevier.com/9780123742391/

ANATOMY OF NEUROPSYCHIATRY

THE NEW ANATOMY OF THE BASAL FOREBRAIN AND ITS IMPLICATIONS FOR NEUROPSYCHIATRIC ILLNESS

ANATOMY OF NEUROPSYCHIATRY

THE NEW ANATOMY OF THE BASAL FOREBRAIN AND ITS IMPLICATIONS FOR NEUROPSYCHIATRIC ILLNESS

Lennart Heimer, M.D.
Professor Emeritus, Department of Neurological Surgery
University of Virginia, Charlottesville, Virginia

Gary W. Van Hoesen, Ph.D.
Professor Emeritus, Departments of Anatomy and Cell Biology
and Neurology
University of Iowa, Iowa City, Iowa

Michael Trimble, M.D.
Professor of Behavioural Neurology
Institute of Neurology, Queen's Square, London

Daniel S. Zahm, Ph.D.
Professor, Department of Pharmacological and Physiological Science
Saint Louis University, St. Louis, Missouri

ELSEVIER

AMSTERDAM • BOSTON • HEIDELBERG • LONDON
NEW YORK • OXFORD • PARIS • SAN DIEGO
SAN FRANCISCO • SINGAPORE • SYDNEY • TOKYO
Academic Press is an imprint of Elsevier

Academic Press is an imprint of Elsevier
30 Corporate Drive, Suite 400, Burlington, MA 01803, USA
525 B Street, Suite 1900, San Diego, California 92101-4495, USA
84 Theobald's Road, London WC1X 8RR, UK

This book is printed on acid-free paper. ∞

Library of Congress Cataloging-in-Publication Data
Anatomy of neuropsychiatry / Lennart Heimer . . . [et al.].
    p. ; cm.
Includes bibliographical references.
ISBN 978-0-12-374239-1 (alk. paper)
1. Neurobehavioral disorders—Physiological aspects.   I. Heimer, Lennart.
[DNLM: 1. Mental Disorders.   2. Brain—anatomy & histology.   3. Brain—physiology.
4. Nervous System Diseases.   5. Neural Pathways—anatomy & histology.   6. Neural
Pathways—physiology. WM 140 A535 2007]
    RC386.2.A48 2007
    616.8—dc22
                                                                    2007031556

British Library Cataloguing-in-Publication Data
A catalogue record for this book is available from the British Library.

ISBN: 978-0-12-374239-1

For information on all Academic Press publications
visit our Web site at www.books.elsevier.com

to Lennart *in memoriam*

TABLE OF CONTENTS

3

The Anatomy of the Basal Forebrain   27

4

The Greater Limbic Lobe   69

5

Cooperation and Competition of Macrosystem Outputs 101

PREFACE

This book emerged in part from a series of workshops on the functional anatomy of the human brain organized by Lennart Heimer. The first "A New Anatomical Framework for Neuropsychiatric Disorders: Systems Analysis and Hands-on Dissection of the Human Brain" took place in 2003 and was repeated during the three succeeding years as one of the workshops offered by Practical Anatomy & Surgical Education, a continuing medical education and community outreach provider attached to the Department of Surgery of the Saint Louis University School of Medicine. The objective of the "New Framework" workshops certainly would seem unique, namely the teaching of functional neuroanatomy to health care professionals in physical and occupational therapy, psychology, psychiatry, neurology, and neurosurgery, all fields where perhaps a fairly detailed knowledge of the functional-anatomical organization of the brain would seem to be a fundmental element. Nevertheless, the workshops invariably booked to capacity, attracting diverse collections of practitioners from all of these disciplines. The special attraction of the workshop seemed to be the promise of hands-on dissection of the human brain. Some workshop participants complained that their training had lacked such an opportunity, but more lamented that, at the time they had been provided with the opportunity, they were intellectually or emotionally unprepared to appreciate its rich rewards.

It was little anticipated by workshop registrants that Lennart Heimer actually meant "*New* Anatomical Framework" literally, or that, during the course of the workshop, many of them would come to appreciate a way of conceptualizing the inter-relationships of the cortical mantle, deep telencephalic nuclei, and descending and ascending systems of the brainstem that differs in significant respects from what they likely were taught during their professional studies. The "new" functional-anatomical concepts have emerged during the past thirty-five years, gradually, but steadily, gaining currency among the basic neuroscience cognoscenti. In contrast, the more conservatively minded clinical disciplines of the workshop partipants have been slower to embrace the new ideas, which is somewhat unfortunate, insofar as the concepts in question reflect a functional-anatomical basis for comprehending the inextricable linkage of mechanisms (if not the mechanisms themselves) subserving cognitive, emotional, and motor components of behavior. Indeed, despite the pervasive resistance of the clinical neurosciences to these contemporary concepts of brain organization, certain clinico-anatomical constructs that have grown out of them, e.g., "parallel, segregated cortico-basal ganglia-thalamocortical circuits," have achieved prominence in contemporary neurological, neurosurgical, and neuropsychiatric thought. It is a telling fact that such diverse and, in the thinking of some, mutually exclusive (or nearly so) clinical disciplines find common ground in such neuroanatomical substrates. Not surprisingly, breaking down the barriers between neurology, neurosurgery, and psychiatry was a consistent theme in Lennart Heimer's writings, voiced perhaps most strongly in "Perestroika in the Basal Forebrain: Opening the Border between Neurology and Psychiatry" (1991). Indeed, it was in the spirit of strenuously contesting the unnatural schisms between these disciplines that Lennart was joined by Michael Trimble, a behavioral neurologist keenly aware of the behavioral side of neurological disorder, and Gary Van Hoesen, a student of the cerebral cortex and its disorders, particularly Alzheimer's disease, to assist him in presenting the workshops. The core faculty was supplemented during the course of the four years that the workshops were done by contributions from Nancy Andreasen, Wayne Drevets, Stefanie Geisler, Stephan Heckers, Andres Lozano, Joel Price, Paul A. Young, and Scott Zahm.

But it is only partially correct to assert that the book owes its existence to the workshops. To the contrary, Lennart Heimer had long worked to integrate functional-anatomical concepts emerging from the research laboratory, which were based on data generated in rodents and monkeys, with human brain organization. Efforts to do this took the form of extensive analyses of histochemically processed postmortem human brain material (see, e.g., Alheid et al., 1990; Heimer et al., 1999; Sakamoto et al., 1999) and innumerable gross dissections of human brains, leading to the two editions of his textbook for medical students, *The Human Brain and Spinal Cord-Functional Neuroanatomy and Dissection Guide*, regarded as among the most lucid available accounts of this difficult subject. As regards the brain dissections, the antecedents of this book harken back to now locally legendary sessions beginning in the late 1970s and continuing through 2006 in which Lennart demonstrated the prosected human brain to neurology and neurosurgery medical residents at the University of Virginia. These teaching sessions led in the 1990s to the production of an acclaimed series of videotaped human brain dissections and ignited a seemingly perpetual process of refinement and re-recording of the dissections, culminating in those available on the companion site for this book, which reflect the work of a consummate neuroanatomist recorded with state of the art technology. Thus, it was the idea to combine the teaching of the "new" functional-anatomical concepts in juxtaposition with classical demonstrations of human brain organization, as revealed by gross dissection, and liberal recourse to clinical correlations that became the glue, or magic, if you will, that for four years held together a program that consistently merited rave reviews from collections of diverse and, we might add, intimidatingly discerning clinicians.

All of us involved in the writing of this book, some with neuroanatomical training, others with clinical insights, came to realize that knowledge of the significant advances in neuroanatomy of especial relevance for understanding behavior had only reached a proportion of those for whom it is relevant for their daily research or practice. This in part reflects on the somewhat esoteric nature of the subject, neuroanatomy abounding in connectional and histochemical complexities far exceeding those found in other anatomical systems, say, e.g., the liver or heart. But it also relates to the relative entrenchment in current clinical textbooks and discussions of the concept of the limbic system, typically still viewed as it was envisaged some fifty years ago. Neuroanatomy, like all disciplines has progressed in many directions, but of relevance here has been the opening up of our understanding of the anatomy of the basal forebrain and the expansion of the concept of the limbic lobe. Further, modularity has given way to circuitry, as systemic relationships, i.e., macrosystems, have been identified within the anatomical complexity. Such advances have arisen not only from new methods of exploring the brain, including new neuroanatomical staining techniques and brain imaging methods in humans, but also from a need to unite our knowledge of brain function and structure with clinical observations. The task is to enable an integration of brain-behavior relationships which not only makes sense but which also has practical application in treating a spectrum of neurological and psychiatric disorders.

The text however originates from a neuroanatomical perspective. This is not a book on neuropsychiatry or biological psychiatry, and we have not elaborated on clinical presentations of various disease states. There are many other books which do that. Instead, it is hoped that the clinical relevance of the new neuroanatomy will emerge naturally from the individual chapters, and that the interested reader will be stimulated to enhance his or her clinical knowledge with the underlying neuroanatomical principles. However, some interesting pointers are provided by way of "Clinical Boxes," as are some of the laboratory methods and basic science issues enlivened by "Basic Science Boxes."

Lennart Heimer is the principal author of Chapters 1–3. Chapters 1 and 2 provide a brief description of the origin and evolution of the concept of the limbic system and some deficiencies attributed to it as a basis for understanding behavior and human neuropsychiatric disorders. Chapter 3 describes the "new" anatomy—an alternative way to conceptualize brain systems subsumed in the more conventional thinking by the limbic system. First, Lennart reveals how the discovery of the striatopallidal relations of the olfactory system leads logically to [1] discrediting the idea of a limbic system-exptrapyramidal system dichotomy and [2] significantly expanding the role of basal ganglia-thalamocortical functions in behavioral synthesis. He then shows that, in functional-anatomical terms, much of the amygdala emulates cortex and that this is quite consistent with the manner in which the great classical neuroanatomists conceived it. He goes on to show that the definitive, highly characteristic histostructural features attributed to the central nucleus of the amygdala are in actuality much more broadly represented in basal forebrain than credited in most contemporary neuroanatomical accounts, although this, too, was recognized by the classical neuroanatomists. Two "new" major constructs to emerge from these considerations, ventral striatopallidum and extended amgdala, together with the cholinergic basal nucleus of Meynert and the septum, are then considered as subcortical output channels for Broca's great limbic lobe, a magnificent concept that, as Gary Van Hoesen reveals in Chapter 4, nevertheless does not escape the keen logic of the new anatomy unaltered. In Chapter 5, Lennart's erstwhile student and long-time collaborator, Scott Zahm, explores the systems and behavioral implications of the new anatomy. Throughout the book and, particularly, in numerous Clinical Boxes, neurological, behavioral, and neuropsychiatric implications of the new anatomy are illuminated by Michael Trimble, who writes in a style that all readers of the book, from beginning students, to seasoned neuroscientists, to jaded clinicians from all of the brain-concerned disciplines, should find to be accessible and rewardingly informative.

Lennart Heimer's "new" anatomy may seem for some to be not so new. Indeed, as Lennart himself liked to point out, the structure of the brain is no different after 1972 than it was before. What occurred instead is that a realization that Lennart had about how the brain is organized led to a series of further discoveries and revisitations to the classical neuroanatomical literature, culminating in a different way of looking at vertebrate brain organization with significant implications for human brain-behavior relationships and neuropsychiatry. No longer can the psychiatric (limbic system) and neurological (extrapyramidal system) be conveniently compartmentalized and relegated to segregated clinical disciplines. To the contrary, the inextricable interwoveness of the cognitive, emotional, visceral, and somatic realms of human experience is clearly reflected in the new anatomy and continues to gain acceptance in clinical medicine. The experimental neuroanatomical basis for this new way of thinking about brain as described herein is presented in elegant counterpoint to the beauty and elegance of the human brain itself, as revealed in the video recorded dissections included with the book, executed and demonstrated by a master of functional neuroanatomy.

Daniel S. Zahm, Ph.D.
Michael Trimble, M.D.
Gary W. Van Hoesen, Ph.D.

ABOUT LENNART HEIMER

This book represents in large part the legacy of its lead author, Lennart Heimer, who, as a young man, tested the waters of art, sport, and engineering and might well have excelled as a professional in any of them. Although it was the study of medicine to which he ultimately committed, even that, as it turned out, was but a threshold over which he passed to enter the realm of experimental neuroanatomy that finally captured and held him for the rest of his life.

While completing his preclinical studies at the University of Gothenberg, Lennart undertook what was supposed to be a brief course of research at the Anatomical Institute with Knut Larsson studying the effects on the sexual behavior of rats of lesions in the preoptic area. Soon finding himself mystified by the diverse, seemingly kaleidoscopic range of behavioral effects produced by such lesions, he deduced that little of value could come of such experiments without some means of comparing different lesions with each other in terms of how much and which brain structures were destroyed. But, then, of course, this kind of information cannot be acquired and, in any event, is of little use unless one knows ahead something about the size, shape, and composition of different brain structures that might be destroyed by the lesions and how they are related to each other by connections. These kinds of considerations were in Lennart's mind when in 1965 he was recruited by W. J. H. Nauta to the Department of Psychology and Brain Science of the Massachusetts Institute of Technology. He had developed what at the time was the most sensitive silver method to map the pathways of nerve fibers in the brain (Fink and Heimer, 1967) and soon, with the aid of this and other rapidly evolving methodologies of the day, formulated revolutionary new concepts described in this book regarding the fundamental organization of brain, particularly as regards to those structures that regulate the emotions and motivation. By 1972 he assumed a professorship at the University of Virginia and as his long and influential career unfolded there, Lennart's thinking evolved ultimately toward what became his most tenaciously held conviction—that the segregation of learned thought about human brain and mental disorders into the neurological and psychiatric reflects an unnatural and counterproductive divide. Perhaps this is made most clear in the 1991 Perestroika paper, written with Jose de Olmos, George Alheid, and Laszlo Záborszky (Heimer et al., 1991), where it is stated in the conclusion that . . .

> "... Maybe the opening of the borders between neurology and psychiatry will eventually lead to a close alliance between the two disciplines? This would make sense from a functional-anatomical point of view. Instead of debating the boundaries of the limbic system, neuroanatomists are increasingly directing their attention to the unique characteristics of individual neuronal circuits or systems that often cross the changing boundaries of the 'limbic system.' To identify anatomical units and neuronal assemblies with similar afferent and efferent connections and a certain uniformity in their intrinsic organization has become a desirable pursuit . . ."

Lennart retired from the research laboratory in 1996, but continued to teach, write, and speak. At the time of his death in March 2007, at the age of 77, he had planned for the coming year a full lecture and workshop schedule spanning several continents, including a fifth installment of the St. Louis workshop. Lennart's scientific legacy will play out as it will in the coming decades, but his impact as an electrifying teacher and mentor is history already. His students and colleagues will always remember Lennart's kind, gentle, and encouraging spirit and, hopefully, his steadfast conviction that knowledge of the neuroanatomical organization of the brain is central to understanding its functions and disorders.

July 2007

ACKNOWLEDGMENTS

The authors are indebted to a number of individuals and organizations without whose support this book could not have been written. For many years, Dr. John Jane, Head of the Department of Neurological Surgery at the University of Virginia was a source of unending encouragement and support to Lennart Heimer in the capacity both of trusted friend and administrator. Dr. Jane's capable assistant, Karen Saulle, provided invaluable administrative and clerical assistance to Dr. Heimer, as did Scott Creasy technical support and Dr. William Ober contributions to the artwork. Gary Van Hoesen received abundant assistance of all sorts, including IT, from Carla Van Hoesen. Paul Reimann at the University of Iowa provided Dr. Van Hoesen with photography support. Scott Zahm was superbly assisted in the laboratory by Beth DeGarmo and Evelyn Williams. Karen Hutsell, Charlotte Ruzicka, Tami Mooseghian, Dave Young, and Paul H. Young, M.D., at Practical Anatomy & Surgical Education helped to make the workshops the successes that they were. The authors gratefully acknowledge support from the University of Virginia School of Medicine (Heimer), University of Iowa School of Medicine (Van Hoesen), Institute of Neurology, Queens Square, London (Trimble), and Saint Louis University School of Medicine (Zahm) and USPHS NIH grants NS-17743 (LH), NS-14944, and NS-19632 (GVH) and NS-23805, DA-15207, and MH-70624 (DSZ). Heartfelt thanks also go to Hakon Heimer, and to Hilary Rowe, Carrie Bolger, and Carl M. Soares at Elsevier.

1

THE LIMBIC SYSTEM; A CONCEPT IN PERPETUAL SEARCH FOR A DEFINITION

1.1 THE BIRTH OF THE LIMBIC SYSTEM

The concept of the limbic system was proposed by Paul MacLean (1949; 1952) based on his own studies of temporal lobe epilepsy and James Papez's proposal that "the hypothalamus, the anterior thalamic nuclei, the gyrus cinguli, the hippocampus and their interconnections constitute a harmonious mechanism which may elaborate the functions of central emotion, as well as participate in emotional expression" (Papez, 1937). Papez's highly speculative theory of a central mechanism of emotion was based on studies by Cannon (1929), Bard (1928), and others, who promoted a central origin of emotion with the hypothalamus as a key structure in the expression of emotions (see Clinical Box 1). Papez supported his theory with evidence from the clinical literature on the symptoms following lesions of various parts of his circuit (known widely as the "Papez circuit"; Fig. 1.1). The inclusion of the hippocampus, for instance, was supported by the observation that Negri bodies, which are part of the pathology of rabies (characterized by intense emotional symptoms),

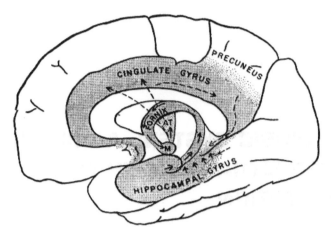

FIGURE 1.1 Diagram of the brain by Paul MacLean (1949) illustrating his concept of the visceral brain (shaded) and the "Papez circuit" (arrows). Abbreviations: M—mammillary body; AT—anterior thalamic nucleus. (Reprinted with permission)

are present in hippocampal pyramidal cells. Papez also found evidence in the clinical literature that the cingulate gyrus is "the seat of dynamic vigilance by which environmental experiences are endowed with an emotional consciousness" (Papez, 1937).

Although neither Papez nor MacLean made reference to Broca in their original 1937 and 1949 papers, it is difficult to invoke the image of the cingulate and parahippocampal gyri without mentioning the great limbic lobe of Broca (1878). On the basis of extensive comparative anatomical observations, Paul Broca, like Thomas Willis (1664) before him, noticed that the cingulate gyrus (callosal gyrus) and parahippocampal gyrus (hippocampal gyrus) form a border (limbus) around the corpus callosum and brainstem. Since primary olfactory input seemed to enter both the cingulate and parahippocampal gyri, the sense of smell appeared to have an especially dominant influence on the functions of this cortical ring.[1] Although hardly ever mentioned in the literature, it is interesting to note that Broca associated the olfactory input with the emotional functions reflecting "the brute within," thus implying a connection between the great limbic lobe and "lower instincts" related

---

[1]The term *rhinencephalon* was sometimes used in the past as a synonym for the limbic lobe, a usage, however, that gradually disappeared in the 20th century, when it was shown that only a restricted part of the limbic lobe is directly related to olfaction (see Chapters 2 and 4).

to emotions that underlie behavior. In hindsight, and with special reference to the continuing evolution of the limbic system concept, it is tempting to agree with Pozzi, who in 1888 wrote that Broca's great limbic lobe was "perhaps Broca's greatest claim to admiration by posterity" (referenced in Schiller, 1979).

Amygdala was not included in Papez's original theory of emotion, but MacLean, with reference to the findings by Klüver and Bucy (1937), made amygdala one of the epicenters in a more extensive system, which he originally called the "visceral brain" (MacLean, 1949). MacLean chose the term *visceral* in the old-fashioned sense of strong, inward feeling (MacLean, 1978), but he soon changed the name to the more neutral term *limbic system* (MacLean, 1952) because of complaints from physiologists, who generally have a more narrow definition of the term *visceral*. MacLean, in choosing the term *limbic*, no doubt recognized the importance of the limbic lobe of Broca as a strikingly characteristic part of the limbic system.

Based on cytoarchitectonic criteria, Yakovlev (1948) developed a functional-anatomical theory that in some aspects is similar to the one developed by MacLean. Yakovlev's studies of the limbic lobe, furthermore, supported the inclusion of the orbitomedial prefrontal cortex and insula in the concept of the limbic system. MacLean, like Papez before him, also made a note of the special relations between the limbic lobe and the precuneus in the context of visceral-emotional, especially sexual, functions (MacLean, 1949). The strategic position of the precuneus in the context of emotional functions and self-awareness is discussed further in Chapter 4.

In his further elaboration of the limbic system concept, MacLean subdivided the limbic system into three main subdivisions related to three nuclear groups (Fig. 1.2): the amygdala, the septum, and the anterior thalamic nuclear complex. He considered these nuclear groups as "hubs for neuronal communication between the limbic cortex and the brainstem." In a broadly sweeping functional analysis extracted from a large number of clinical, physiological, and behavioral studies, he related these subdivisions of the limbic system to self-preservation (the amygdala division), preservation of the species (the septal division), and family-related, including maternal, behavior (the thalamo-cingulate division). For further discussion of the functional aspects of these subdivisions, the reader is referred to MacLean's book, the *Triune Brain in Evolution* (1990), in which the limbic system is placed in an evolutionary context (see Chapter 2).

Since its inception in the middle of the last century, the limbic system concept has been the preeminent theory for explaining emotional behavior, and it has had a great impact in many fields of neuroscience, especially psychology and psychiatry. The urge to explain the biology of emotional functions and psychiatric disorders propelled the limbic system forward to become a highly popular neuroscientific theory.

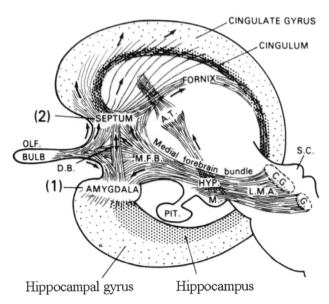

FIGURE 1.2 Diagram by Paul MacLean (1958) placing emphasis on the medial forebrain bundle (MFB) as a major line of communication between the limbic lobe and the hypothalamus and midbrain. Major components of MacLeans's concept of the limbic system, including the septum, amygdala, and anterior thalamus (A.T.), are also indicated. 1 and 2 denote ascending pathways to limbic lobe, with emphasis on divergence of fibers from the medial forebrain bundle (MFB) to the amygdala and septum. Addtional abbreviations: C.G.—central gray of midbrain; G.—ventral and dorsal tegmenal nucleus of Gudden; D.B.—diagonal band; Hyp.—hypothalamus; L.M.A.—limbic midbrain area of Nauta; M.—mammillary body; Olf. bulb—olfactory bulb; Pit.—pituitary gland; S.C.—superior colliculus. (Reprinted with permission)

CLINICAL BOX 1

History of the Limbic System

The historical importance of the limbic system concept, ill-defined and confused as it is, should not be underestimated. It has been known for centuries that people suffer from what may loosely be referred to as emotional disorders, with many suggested potential causes that differ, depending on the historical epoch that is chosen. Most famous and relevant to this discussion is a statement by the Greek philosopher-physician Hippocrates that "from the brain come joys, delights, laughter and sports, and sorrows, griefs, despondency and lamentations. . . . and by the same organ we become mad and delirious, and terrors and fears assail us" (Adams, 1939). At the time of Hippocrates' writing, many disorders, from epilepsy to madness, were thought to be the result of divine influence. Hippocrates opined that they are essentially somatic in origin and that the right approach to understanding them is through the natural sciences.

Although great strides in neuroanatomy and neurophysiology were made in the 19th century, there was little progress in understanding how emotions are represented neurologically. At the end of the century, William James suggested that the emotions are derived from peripheral sensory inputs to the brain, which activate motor outputs—and the resulting bodily sensations are perceived as emotion (James, 1884). As discussed in the book *The Emotions* (James and Lange, 1922), the Danish scientist C. G. Lange stated a similar hypothesis in 1885; therefore, the theory is often referred to as the James-Lange theory. In a nutshell, we do not run away from something because we are frightened, but we experience fear because we are running away. However, there was no obvious cerebral location for the generation of the emotions, although the sensory experiences were known to be received in the parietal cortex. The James-Lange hypothesis was soon tested and appeared to be wrong on two counts. First, it was demonstrated in animals that removal of the cerebral cortex on both sides did not abolish the expression of emotion (Vanderwolf et al., 1978; Whishaw, 1990). Further, it was revealed that stimulation of various structures buried deep within the brain could lead to the release of emotion. These observations formed the basis for a neuroscience revolution. That certain brain structures could form the foundation of an emotional brain system was a stunning conceptual departure for neurology and became a launch pad for the discipline of behavioral neurology—that is, that branch of neurology that tries to understand how the brain modulates and relates to behavior and neuropsychiatry.

Apropos is the Jamesian or "bodily feedback" theory, it may yet contain a kernel of truth (Damasio, 1996; LeDoux, 1996). Sensory stimuli from various body parts involved in emotional expressions are no doubt closely integrated with brain structures and circuits involved in emotional functions. In fact, it may not be too far-fetched to paraphrase Lautin (2001): "Peripheral (James) and central (Papez-MacLean) theories of emotions are two sides of the same coin." One of the criticisms of MacLean's limbic system concept or central theory of emotion was that the key structures involved seemed relatively isolated from the isocortex. It was as if the rider (the neocortical mantle) lacked reins with which to control the horse (the limbic system). This seemed to imply some self-contained arrangement, and if there was a neurology to psychiatry, then it all rested on an understanding of the limbic system. It should be recalled that MacLean originally envisioned the connections between the neocortex and the limbic lobe to be limited in scope (MacLean, 1955). Now we know that they are indeed massive. In all fairness, however, MacLean developed his theory 50 years ago. Classic tract-tracing studies in the late 1960s (Pandya and Kuypers, 1969; Jones and Powell, 1970) outlined a large system of cascading pathways from primary cortical sensory areas via unimodal association areas to multimodal regions in prefrontal and temporal association cortex, which in turn is closely related to both the hippocampal formation and the amygdala. More recent anatomical studies, furthermore, have shown significant connections between the limbic lobe and most surrounding neocortical areas, including the frontal lobe (Van Hoesen et al., 1993).

1.2 THE CONTINUING EVOLUTION OF THE LIMBIC SYSTEM

1.2.1 The "Limbic Midbrain Area" of Nauta

Based on clinical and physiologic data, MacLean (1955) had already further expanded the limbic system by including the midbrain central gray and surrounding parts of the reticular formation. Nauta amplified MacLean's physiologic observations on limbic system–brainstem relations by describing direct and indirect hippocampal projections to cell groups in the midbrain, including ventral tegmental area, mesencephalic central gray, dorsal and median raphe nuclei, and other cell groups in mesencephalic tegmentum, including the dorsal and ventral tegmental nuclei of Gudden (Nauta, 1958), which, collectively, have come to be referred to as Nauta's "limbic midbrain area."[2]

These pioneering neuroanatomical tract-tracing studies formed the basis for a significant expansion of the limbic system concept. Central to Nauta's theory are direct and indirect, reciprocal connections between the limbic forebrain (e.g., orbitofrontal cortex, cingulate gyrus, hippocampus, amygdala, and septum) and his limbic midbrain area (Fig. 1.3). These pathways form the basis for what is usually referred to as a "limbic forebrain-midbrain circuit" or "limbic forebrain-midbrain continuum," which is characterized by powerful direct and indirect connections through the hypothalamus. Based on Nauta's contributions, the idea emerged of a "distributed limbic system" with links to frontal neocortical regions, basal ganglia, and brainstem (Nauta, 1986).

1.2.2 "The Greater Limbic System" of Nieuwenhuys

In typical valorous fashion, Nieuwenhuys has recently expanded the limbic system significantly by adding extensive medial and lateral tegmental areas throughout the rostrocaudal extent of the brainstem (Nieuwenhuys, 1996). Included in the "greater limbic system" (Fig. 1.4) are monoaminergic cell groups—although dopaminergic cell groups in the substantia nigra and ventral tegmental area are apparently not included[3]—and a number of brainstem

---

[2]The Gudden nuclei, although well defined in most mammals, are not prominent in the primate, including the human, where especially the ventral tegmental nucleus of Gudden appears to have lost its individuality in the process of becoming an integral part of the reticular formation (Hayakawa and Zyo, 1983). Based on studies in animals, the Gudden nuclei have long been known for their close relation to the mammillary body, which in turn receives input from the hippocampal formation via fornix.

[3]It will be impossible to devote to the midbrain dopaminergic neurons the attention they deserve in relation to the main themes of this book. Suffice it to say that, while dopamine ranks among the most studied of neuroactive molecules, its biological actions remain among the least well

Continued

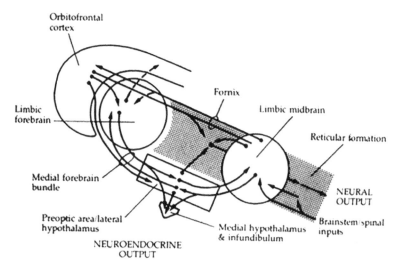

FIGURE 1.3 Modified from a diagram by Walle J. H. Nauta (Nauta and Haymaker, 1969) illustrating proposed connectional relationships between the limbic forebrain and limbic midbrain areas in relation to motor and neuroendocrine output. (With permission)

Continued

understood. After Carlsson et al. (1958) demonstrated that it is dopamine, rather than norepinephrine, levels that correlate with restored motor function in the reserpinized (depletes norepinephrine, epinephrine, dopamine, and serotonin) rabbit brain, DOPA (catecholamine precursor) was shown to restore motor function in Parkinson's disease patients (Sano, 1960 [Japan, DL-DOPA, i.v.]; Birkmayer and Hornykiewicz, 1961 [Germany, L-DOPA, i.v.]; Barbeau, 1962 [Canada, L-DOPA, oral, first reported in Geneva in 1961]. This action was thought to be related to replacement of dopamine, which Ehringer and Hornykiewicz (1960) had shown was depleted in Parkinsonian brains (see Carlsson, 2002 for historical details). Subsequently, Carlsson and Lindqvist (1963) demonstrated that the antipsychotic actions of chlorpromazine and haloperidol reflect the capacity of these drugs to antagonize dopamine receptors, indicating that dopaminergic function transcends motor control and issuing in the era of the dopamine hypothesis of schizophrenia. In the meantime the great Scandanavian histochemists were gearing up to map the positions of the catechoaminergic neurons (Dahlstrom and Fuxe, 1964) and fibers (Anden et al., 1964, 1965, 1966a, 1966b) throughout the neuraxis, and it soon was realized that cell groups embedded in the brainstem reticular formation provide massive ascending and descending catecholaminergic and indoleaminergic projections to the telencephalon and spinal cord, respectively. The dopaminergic projection systems were categorized as mesostriatal, from neurons located in the substantia nigra pars compacta to terminations in the caudate nucleus and putamen, and mesocorticolimbic, from neurons in the mesencephalic ventral tegmental area to terminations in a variety of structures associated with limbic system, including the prefrontal and other "limbic" cortices, septum, accumbens, bed nucleus of stria terminalis, preoptic region, lateral hypothalamus, amygdala, hippocampus, and habenula. Among the dopaminergic mesolimbic projections, the most attention by far has been given to the mesoaccumbens axis, which has been implicated in the regulation of locomotor activation (e.g., Kelly et al., 1975), reward (Wise, 1978), motivation (Salamone et al., 1994), and memory and learning (Wise, 2004), and decades ago was identified as the primary focus of attack by drugs of abuse to "hijack" the reward system (Wise, 1980; see also Nestler, 2005).

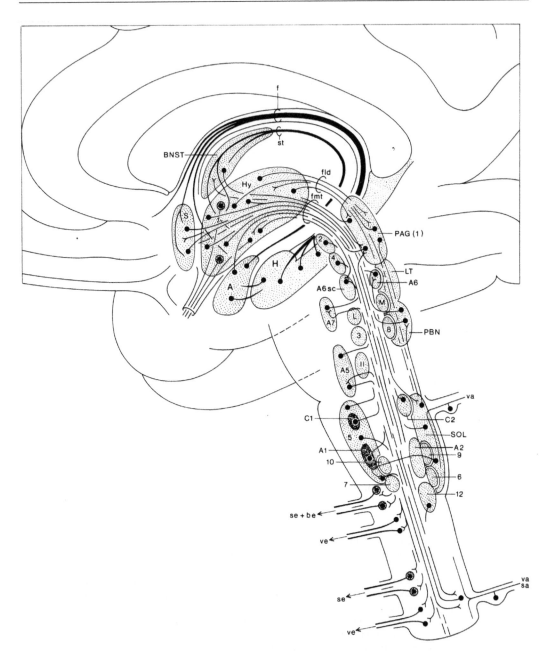

FIGURE 1.4 Limbic circuitry according to Nieuwenhuys (1996), who continued the expansion of the limbic system into the brainstem begun by Nauta. The reader is referred to the original text for meanings of abbreviations. (Reprinted with permission)

regions from which behavioral and autonomic responses can be elicited. Structures like the vagal-solitary complex, parabrachial nucleus, and other regions in the mesopontine tegmentum are added to the limbic system. In contrast to surrounding classic myelinated sensory and motor pathways, the limbic brainstem regions are characterized primarily by thin, unmyelinated fibers and correspond to what Nieuwenhuys and his colleagues have defined as the "core of the neuraxis." Many of these limbic components comprise countless varicose, unmyelinated fibers that are enriched in neuropeptides, which, according to Nieuwenhuys, signifies nonsynaptic secretion of neuromediators.

In summarizing Nieuwenhuys's contribution to the limbic system concept (Nieuwenhuys et al., 1998), it is probably fair to say that he followed in the footsteps of Nauta to continue the caudal expansion of the limbic system by including many additional brainstem structures (except the classic sensory and motor systems) known to be of importance to integrated adaptive responses. From a clinical perspective, this made perfect sense (see Clinical Box 2). Whereas amygdala, hippocampus, and hypothalamus are still part of Nieuwenhuys's greater limbic system, the cingulate and parahippocampal gyri—that is, the original limbic lobe of Broca—is considered in his scheme to be an interface between the greater limbic system and the neocortex. The ventral striatum (limbic striatum) is conceived of as an interface between the limbic system and the extrapyramidal motor system (see Chapter 3). Holstege's (1992) "emotional motor system," which is discussed in Chapter 5, is an integral part of Nieuwenhuys's "greater limbic system."

1.3 WHY ISN'T THE CEREBELLUM AN INTEGRAL PART OF THE LIMBIC SYSTEM?

The cerebellum is reciprocally connected with the telencephalon, and the massive cerebro-cerebellar connections have long been known to be crucially important for motor functions. Indeed, throughout most of the 20th century, the cerebellum was generally considered to be important primarily, if not exclusively, for control of movements, muscle tone, and postural adjustments. The realization—actually, the rediscovery and popularization (Schmahmann, 1997b)—that the cerebellum is involved not only in sensorimotor and cognitive functions but in the regulation of emotional behavior invites the temptation to define a "limbic" cerebellum. Indeed, neuroanatomical tract-tracing studies have shown that the cerebellum has robust direct connections with the hypothalamus and many parts of the brainstem, including the "limbic midbrain area" (Haines, 1997; Snider, 1976). Apparent cerebellar involvement in all aspects of human behavior is based on a rich network of neuronal circuits including cerebro-cerebellar feed-forward and feed-back loops and a multitude of connections between the cerebellum and brainstem.

As reviewed by Schmahmann (1997a), recognition of cerebellar involvement in emotional and cognitive functions has a long history, which largely had been forgotten, or at least ignored, by the neuroscience community, including most textbook authors (see Clinical Box 2). Thus, the recent, more general acceptance of a role for the cerebellum in all aspects of behavior may seem tantamount to a paradigm shift in our understanding of brain function, but it just reflects an exercise in group recollection. According to Schmahmann (1997a), it is tempting to draw a parallel between recent developments in our understanding of the functions of the cerebellum and those of basal ganglia, which also used to be considered in a purely motoric context, as reflected in the unfortunately still common evocation of a "limbic system-basal ganglia" or "limbic system-extrapyramidal" dichotomy (see Chapter 2). However, the reasons for changing our thinking about the functions of these two major brain structures differ. The discovery of the ventral striatopallidal system and associated experimental neuroanatomical demonstration that the interconnections of the ventral striatum, ventral pallidum, mediodorsal thalamus, and prefrontal cortex form a cortical-basal ganglia-thalamocortical "loop" circuit seeded the concept of cortico-subcortical reentrant circuits that has come to dominate thought on basal ganglia function (see Chapter 3). In contrast, the apparent "paradigm shift" in thinking with regard to the functions of cerebellum is more like a recovery by the neuroscience community from a collective "neglect syndrome" (see Clinical Box 2).

1.4 THE EVOLUTION OF THE LIMBIC SYSTEM: NO END IN SIGHT

Maybe one of the problems with the concept of the limbic system as an emotional system can be traced back to its very beginning when Papez attempted to attribute emotional functions to a number of closely related anatomical structures with close relation to the hypothalamus (including the mammillary body) but with hardly any experimental evidence that any of them, except the

CLINICAL BOX 2

Cerebellum and the Limbic System from a Clinical Perspective

The book edited by Schmahmann (1997b) enlightened the general neuroscience community to the importance of cerebellum as a regulator not only of motor function but also cognitive and emotional aspects of behavior. The relationship between the cerebellum and cognitive-affective functions is now summarized in the so-called "cerebellar cognitive affective syndrome." Appreciation of this condition, based on clinical observations of cerebellar disorders and on experimental animal studies, has a long history, which prominently involves neuropsychiatric disorders. Following the introduction of the limbic system in the second half of the 20th century, the role of the cerebellum in emotional functions was conveniently explained on the basis of a close relationship between the cerebellum and a rapidly expanding limbic system.

The title "Rediscovery of an Early Concept" sets the stage for a revealing discussion of nonmotor cerebellar functions (Schmahmann, 1997a). For close to 200 years, cognitive and affective symptoms have repeatedly been attributed to cerebellar pathology, but since many of the early case studies and clinical reports lacked the rigor of modern scientific presentations, they were often looked upon as anecdotal accounts and generally ignored. Equally puzzling is the fact that autonomic and affective responses revealed in many experimental animal studies from the second half of the 20th century, including especially electrical stimulations of cerebellar structures, suffered neglect in the shadow of more conspicuous motor effects. In particular, stimulations and electrode recordings in patients with emotional disorders in the late 1960s and early 1970s (Nashold et al., 1969; Heath et al., 1974) provided significant evidence for the involvement of cerebellum in emotional functions and neuropsychiatric disorders. These studies foreshadowed the current interest in structural and functional neuroimaging studies that show unequivocal cerebellar involvement in cognitive and affective functions, as well as in neuropsychiatric disorders, including both schizophrenia (Nopoulos et al., 1999) and affective disorders (DelBello et al., 1999).

Incidentally, Schmahmann's original suggestion that the effects of cerebellar lesions, characterized by erratic modulation of cognition and poor cognitive performance, be referred to as "dysmetria of thought" (Schmahmann, 1991)— that is, "cognitive dysmetria"—is currently being promoted as an integrative theory to explain the broad range of symptoms characteristic of schizophrenia (Andreasen et al., 1998). With the popularity of the limbic system at a peak in the second half of the 20th century, clinicians predictably pointed to functionally relevant connections between the cerebellum and an increasingly distributed limbic system, which, as reviewed earlier, eventually includes a centrally located continuum from the rostral forebrain (including the limbic lobe, septum, hippocampus, and amygdala) to structures in the caudal brainstem (including autonomic nuclei, monoaminergic systems, and the reticular formation). A certain gross topography of cerebellar function was also suggested based on classical anatomy and correlated functional-anatomical studies in the sense that the vermis and fastigial nucleus were regarded as the main representatives of a "limbic cerebellum," whereas cognitive functions were thought to be more closely related to the cerebellar hemispheres. An increasing number and complexity of connections between cerebellum and various parts of the greater limbic system will undoubtedly be revealed in coming years, and they will serve to emphasize a role for cerebellum in emotional functions and neuropsychiatric disorders. It will be important to find out what exactly cerebellum contributes to various emotional functions and adaptive behaviors.

hypothalamus, is related to emotional functions. The hippocampus thus became a founding member of the club before it was shown that the amygdala, rather than the hippocampus, is the key telencephalic structure in the context of emotional functions. The hippocampus, on the other hand, is key to the memory system. The distinction, however, may no longer be a problem, since with the gradual expansion of the limbic system, definitions of its functions have broadened considerably. Representative reflections of this are apparent in statements such as those posing the limbic system—for example, as "a determinant of the organism's attitude toward its environment" (Nauta and Feirtag, 1986) or declaring that its main function (singular) is "adaptation, including homeostasis, response to physiological or behavioral demand, and reproduction" (Herbert, 1997). Considering the vagueness and ambiguity in these "statements of purpose," any system, limbic or otherwise, charged with carrying out these duties clearly has a "full plate," as does anybody who might try to describe the system. The anatomical characterization of such a comprehensive functional system would be a difficult task to say the least. In fact, it may be impossible without enlisting practically the entire brain, including the cerebellum! Nonetheless, it has been universally presumed that this is what the limbic system does, and during most of the 20th century, the fame and influence of the limbic system have continued unabated both in basic and clinical neuroscience, despite the problems created by its perpetual expansion and mobile boundaries. That there seems to be no end in sight to the continuing evolution or resolution of the limbic system is perhaps best illustrated with the example of LAMP, the limbic system-associated membrane protein, which, upon having been proposed as a solution to the dilemma of defining the limbic system (Reinoso et al., 1996), was found to be present in many presumably nonlimbic parts of the nervous system.

Ironically, to the extent that the continuing expansion of the limbic system has been fueled by neuroanatomical advances, experimental neuroanatomy may well hold the seeds of its demise. The rise of contemporary experimental neuroanatomical and histotechnical methods in the 1960s, 1970s, and 1980s and the combination of these methods with molecular biological approaches in the 1990s and 2000s have fueled an acquisition of exponentially increasingly amounts of detailed information about neuronal pathways and circuits. Based on these advances, neuroscientists have begun, little by little, to piece together "an anatomy of partial functions" (LeDoux, 1996; Morgane et al., 2005). It is a fair guess that this process will continue until we have obtained a reasonably good understanding of how different neuronal systems and circuits of the brain contribute to various aspects of emotional functions and rational adaptive behavior. The fallout of these advances has been a more concerted effort directed toward exposing the shortcomings of the limbic system, which, in turn, has unleashed a stiff counter offensive by its proponents (Chapter 2).

To further develop a rationale for abandoning the limbic system concept as a guide for the scientific exploration of emotional functions, the next chapter will summarize some of the controversy surrounding MacLean's concept of a "triune brain," of which the limbic system is an integral part. Having questioned the relevance of the limbic system concept in modern neuroscience research, we will endorse an alternative solution based on the study of specific emotions. In order to facilitate this endeavor, we will summarize recent anatomical studies that have resulted in a new way to look at the basal forebrain (Chapter 3), and we will present a modified anatomical version of Broca's original limbic lobe (Chapter 4), culminating in the presentation in Chapter 5 of a systems-based anatomical framework for the study of emotional functions and neuropsychiatric disorders.

In addition to MacLean's opus on the triune brain, a large number of books (Adey and Tokizane, 1967; DiCara, 1974; Livingston and Hornykiewicz, 1978; Isaacson, 1982; Koella and Trimble, 1982; Trimble and Zarifian, 1984; Doane and Livingston, 1986; Gloor, 1997; Lautin, 2001) and an even larger number of commentaries and review papers, many of which are referenced in Kötter and Meyer (1992) and Morgane et al. (2005), have been written on the subject of the limbic system. There also have been attempts to make sense of the vast, heterogeneous collection of individual opinions about the limbic system (Kötter and Meyer, 1992; Anthoney, 1994) and some additional excellent papers provide historic perspectives and analyze the appeal of the limbic system (Durant, 1985; Lautin, 2001).

2

THE ERODING RELEVANCE OF THE LIMBIC SYSTEM

During the 20th century, the influence of the limbic system increased both in basic and clinical neuroscience, despite the problems created by its gradual expansion and diffuse boundaries. Then, in 1990, MacLean published his opus magnum, *The Triune Brain in Evolution*, or "three brains in one," in which the limbic system is placed in an evolutionary context. Since its publication, the theory of the triune brain has been often debated. It has been referred to by Durant as "more a metaphor than a theory" (in Harrington, 1991), and it has also been said to be "by far the best concept we have for linking neuroscience with the social sciences" (Cory, 2002).

2.1 THE TRIUNE BRAIN CONCEPT AND THE CONTROVERSY SURROUNDING IT

MacLean suggested that a paleomammalian brain (represented by the limbic system, which according to MacLean is important in emotional behavior related to feeding, reproduction, and parental functions) was added in early mammals to the already existing reptilian brain (the basal ganglia, important for daily routines and ritual displays related to aggression, territoriality, and courtship). The evolutionary process was later followed in modern mammals by the development of the neomammalian brain (neocortex, related to problem solving and other cognitive functions). MacLean's notion of a hierarchical structure represented by the triune brain had popular appeal, and it reached beyond the research community as reflected by the books *The Ghost in the Machine* (Koestler, 1967) and *The Dragons of Eden* (Sagan, 1978), in which MacLean's notion of the triune brain was used in speculations on the human condition and destiny.

Almost as soon as MacLean's triune brain was proposed, it came under fire. Several reviewers of the concept questioned its compatibility with modern views of evolutionary neurobiology (Durant, 1985; LeDoux, 1991; Butler and Hodos, 1996; Reiner, 1997). As it turned out, while MacLean was developing his idea in the late 1960s and early 70s (MacLean, 1970), the field of comparative neuroanatomy was undergoing a revolution of its own. As is usually the case in neuroscience, the revolution was triggered by improved technology, in this case, histotechnical advances heralding the emergence of new experimental methods to trace neuronal connections (see Basic Science Box 1). Application of these new techniques to a number of mammalian and nonmammalian species in the 1960s and 1970s revealed that the blueprint of forebrain organization is similar in all vertebrates (Karten, 1969; Nauta and Karten, 1970; Ebbesson, 1980; Butler and Hodos, 1996). In other words, homologues of MacLean's limbic system are present in nonmammalian vertebrates. Northcutt and Kaas (1995) have made the point forcefully that it becomes more difficult to talk about new and old parts of the telencephalon if the fundamental plan of the forebrain is similar in all vertebrates.

Another aspect of forebrain organization brought to light as a result of histotechnical advances in the late 1960s and early 1970s essentially dismissed the then widely accepted notion that practically the entire telencephalon of primitive vertebrates is devoted to olfaction. This idea was laid to rest when it was shown in several such species that only a part of telencephalon receives direct olfactory bulb input (Scalia et al., 1968; Heimer, 1969; Ebbesson and Heimer, 1970), a realization that was soon followed by the discovery that other sensory systems—for example, the visual system (Ebbesson and Schroeder, 1971; Cohen et al., 1973)—as well as auditory and somatosensory systems also have telencephalic representations in nonmammalian vertebrates. To reiterate, the nonolfactory part of the telencephalon of mammals and nonmammalian vertebrates contains specific areas for auditory, visual, and somatosensory input. In other words, this fundamental and important aspect of forebrain organization is similar in all vertebrates.

In summarizing the meaning of these studies, it seems safe to conclude that MacLean's idea of a successive development of reptilian, paleomammalian, and neomammalian parts of the brain, as reflected in the notion of a triune brain, is incorrect. All vertebrates have multiple sensory telencephalic areas in

BASIC SCIENCE BOX 1

Neuroanatomical Methods

Knowledge of the pathways in the brain and spinal cord cannot in itself explain how the nervous system works, but the anatomic and chemical mapping of neuronal circuits and synaptic relations is a prerequisite for such understanding. Fortunately, the neuroscientist in the pursuit of such information now has a large number of sophisticated methods available to produce highly specific labeling of neuroanatomical structure in sections through the brain that can be viewed at light and electron microscopic levels of resolution.

Although individual axons can be traced by the famous Golgi method (Golgi, 1873) and with the aid of intracellular injection techniques, the tracing of long axonal projections is usually performed much more efficiently with other techniques. The modern era of tract-tracing began in the 1950s when the Dutch neuroanatomist Walle Nauta and the Swiss chemist Paul Gygax developed the so-called "suppressive" Nauta-Gygax silver method (Nauta and Gygax, 1954) for tracing degenerating axons in experimental animals following transection of a pathway or destruction of the area where the pathway originates. The reference to the word *suppressive* denotes the ability of the method to suppress the staining of normal fibers, which made it easy to follow the course of the degenerating fibers as they meandered through the brain and spinal cord. An important reason for the success of the silver method was the fact that its most sensitive modifications, which became available at the end of the 1960s (Fink and Heimer, 1967; de Olmos, 1969), made it possible to trace pathways throughout the central nervous system in order to visualize the entire axonal projections including their terminal end structures (boutons) at the light microscopic level. The areas of terminations could then be confirmed and investigated further with the electron microscope, which was by then a standard piece of equipment in most anatomy laboratories. In Chapter 3 we will explain how the silver methods were instrumental for the discovery of the ventral striatopallidal system and extended amygdala, which are major parts of the new anatomical framework promoted in this book.

In the 1960s another histotechnical breakthrough was achieved with the development of the Falck-Hillarp fluorescence histochemical method for the tracing of monoamine neurons in the central nervous system (Falck et al., 1962). The ability to characterize a neuron by its transmitter was a milestone on the road toward the establishment of a subdiscipline of chemical neuroanatomy, which mushroomed into a huge and sophisticated field in its own right, especially following the introduction of immunohistochemical methods (Sternberger, 1979). The field of chemical neuroanatomy features its own scientific journal and a *Handbook of Chemical Neuroanatomy*, with more than 20 volumes published so far. As reflected in many parts of this book, it is difficult to overestimate the importance of chemical neuroanatomy in the field of neuropsychiatry, especially since the methods used are often applicable not only to experimental animal brains but also to postmortem human brains.

In the 1970s, anterograde and retrograde labeling techniques replaced the silver staining as the methods of choice in experimental tract-tracing. These methods exploit the biological process of axonal transport following injection of a tracer in the area either of origin or termination of a pathway. In antero-grade tracing, radioactively labeled amino acids, neurobiotin, or biotinylated dextran amine (BDA), and the plant lectins wheat germ aggluttinin and Phaseolus vulgaris-leucoagglutinin (PHA-L) can be used to trace the pathway to its termination, whereas retrograde labeling utilizes horseradish peroxidase (HRP), the β subunit of cholera toxin, various fluorescent tracers or BDA to identify the neurons of origin of a pathway following an injection of the tracer into the terminal field. Successive links in neuronal chains can be identified by using combinations of methods at the light- and electron microscopic level. Multineuronal pathways can also be traced using axonal transport of live viruses. The major advantages of using live viruses, usually herpes simplex or pseudora-bies virus, are twofold. First, they are translocated by axonal transport, as are all axonal tracers, but more important, they also are transferred from one neuron to another, apparently preferentially at synaptic contacts. Second, they replicate after the transneuronal transfer, which in turn amplifies the "tracer signal" in the recipient neuron, and this can be detected with immunohistochemical methods. The ability to trace functionally relevant multisynaptic connections makes the viral transport method ideal for tracing cortico-subcortical reentrant circuits (Middleton and Strick, 2001), the elucidation of which has profoundly influenced thought on the functional-anatomical basis of neuropsychiatric dis-orders (see "The Notion of Parallel Cortico-Subcortical Reentrant Circuits" in Chapter 3). The viral tracing method, however, has not gained widespread popularity. One reason for this is that the method is technically demanding, but it seems, rather, that many investigators have shied away because neurotropic viruses cause infection in the animals and may also be a health hazard to humans if not properly handled. Nonetheless, viral tracing carried out with strict bio-safety controls has become a routine procedure in some laboratories.

Recent advances in experimental tract-tracing include sophisticated combina-tions of anatomical, physiological and histochemical techniques. Molecular biological methods can be used to study gene expression in anatomically and physiologically identified neurons. The difficult problem of studying connections in the human brain has received an unexpected boost from the development of diffusion tensor imaging (DTI), which is based on the tendency of water mole-cules to diffuse in the direction of myelinated fiber bundles. Results obtained with DTI in the living human brain by so called "in vivo tractography" have been spectacular, and the method promises to be of great importance in clinical neuroscience. For more information on these and other techniques, which have resulted from the histotechnical revolution of the last quarter century, the reader is referred to a series of three handbooks: "Neuroanatomical Tract-Tracing" (Heimer and RoBards, 1981; Heimer and Záborszky, 1989; Záborszky et al., 2006).

one form or another, similar to that seen in modern mammals. The distinction between "old" and "new" cortex was clearly in a state of evolution (Nieuwenhuys and Meek, 1990; Smeets, 1990) when MacLean published his book on the triune brain. Possibly, in anticipation of these critiques, which seemed an unavoidable response to his book, MacLean preemptively clarified his position. For instance, on page 9 of *The Triune Brain in Evolution*, he mentions in reference to the idea of mammals being under the control of three autonomous brains that this was not what he meant by the notion of the triune brain. Rather, the three brains are "intermeshing," although they are still able to operate "somewhat independently."

Other writers, especially in the fields of psychology, psychiatry, and social science, rose to defend MacLean's construct (Cory, 2002; Gardner, 2002; Panksepp, 2002). For example, one supporter contended that not enough attention has been paid to the fact that MacLean's ideas by nature are subject to "additional developments and refinements" (Panksepp, 2002). Another viewpoint commonly expressed by proponents of MacLean's concepts is that those who have criticized the triune brain and limbic system theories have not offered alternative global theories or "higher-level generalizations" of emotional functions (Cory, 2002).

Although the debate surrounding MacLean's triune brain concept is unlikely to subside anytime soon, it is difficult to deny that his evolutionary neuroethology has had a significant impact in the field of behavioral science. A recurring theme in the defense of MacLean's triune brain concept, limited in some aspects as it may be, is that it nonetheless has explanatory and guiding power in social science, neuropsychiatry, and psychiatry (Cory, 2002; Gardner, 2002; Price, 2002; Ploog, 2003). Furthermore, it is difficult to read the *The Triune Brain in Evolution* without being impressed by MacLean's scholarship and penetrating analysis of the relationship of the brain to adaptive behavior and human nature, based as they are on detailed studies and readings of comparative anatomy, physiology, psychiatry, and clinical neurology.

2.2 THE LIMBIC SYSTEM AND THE CONTROVERSY SURROUNDING IT

For anybody familiar with the continuing evolution of the limbic system concept, it should come as no surprise that an assignment to teach a class or give a lecture on the limbic system always provokes some difficult to answer questions: What does it include? What does it do? As mentioned in Chapter 1, diverse opinions abound regarding what the limbic system includes and does, but all versions of the limbic system are variations on MacLean's original theme summarized in Fig. 1.2.

Already in his first paper, MacLean (1949) added the septum and the amygdala to Papez's original system of emotion, in which the hippocampus formation and the cingulate gyrus were the key telencephalic structures. But it has long been known that the amygdala and the hippocampus have to a large extent quite different functional correlates. Whereas the amygdala is well known for its involvement in emotional aspects of behavior (Klüver and Bucy, 1937; Weiskrantz, 1956), the hippocampus is recognized for its memory functions (Bechterew, 1900; Scoville and Milner, 1957). The posterior hippocampus, in particular, is primarily involved in memory-related cognitive functions rather than in emotional and autonomic aspects of behavior (see Chapter 4). Although MacLean paid close attention to the various interoceptive and exteroceptive inputs driving the limbic system, it is interesting to note that even in *The Triune Brain in Evolution* he makes no reference to classic studies by Pandya and Kuypers (1969) and Jones and Powell (1970) on cortico-cortical pathways (referred to in Clinical Box 1 and Chapter 4), which set out some of the basic tenets that would come to underlie the vast body of memory-related research on the hippocampus.

Another prominent and consistent theme cites the close relation of the limbic system to the hypothalamus. In fact, hypothalamus is the central structure in most people's concept of the limbic system, as reflected in Figures 1.2–1.4 (Chapter 1), which depict some of the best-known renditions of the limbic system. A relationship to the hypothalamus became a defining characteristic of a "typical" limbic structure, such as the septum, amygdala, and allocortex, including the hippocampus and, although this would later prove to be wrong, the olfactory cortex (see Chapter 3). The fact that most of neocortex (isocortex[1]) was known to project to basal ganglia but not hypothalamus led to an important perceived distinction between the limbic system and the basal ganglia. "Limbic versus basal ganglia" and "limbic versus extrapyramidal" were common expressions in both basic and clinical neuroscience throughout the second half of the last century. These phrases became metaphors for a more or less consensus, albeit inaccurate (see Chapter 3), conceptualization of the functional-anatomical organization of the forebrain (see Clinical Box 3).

Since a number of different limbic systems are in circulation (Anthoney, 1994), it follows that what the limbic system includes and does have always to some extent depended on the preferences and interests of its describers. This being the case, it is understandable that the limbic system kept expanding as more and more structures with close connections to the "original" limbic system components were described. Already in 1969, Brodal suggested that the limbic system is "on its way to including all brain regions and functions."

---

[1]Regarding these synonymous terms, the authors favor the use of "isocortex," not because they think of the cortex as homogeneous or uniform, but because all of the cortical areas included in this category have six well-defined layers. The prefix "neo-" carries implications regarding evolutionary chronology that are not necessarily supported by the available data.

CLINICAL BOX 3

The Problem with the Limbic System

The "limbic vs. extrapyramidal dichotomy" has tended to distort our under-standing of the clinical relevance of the limbic system. Suggestions that there are very few direct cortical projections to the hypothalamus and that the hypothala-mus is to be regarded as the principal subcortical projection of the limbic system led to an interesting but rather damaging conclusion, with implications not only for understanding brain-behavior relationships but also for the developing fields of behavioral neurology and neuropsychiatry. Thus, a half century ago, those few neurologists who might reluctantly have conceded that there is an underlying neurology of behavioral expression could be satisfied with a limbic system-hypo-thalamic axis as an explanation for it. In contrast to such a rudimentary neurol-ogy of emotions (and hence psychiatry), the neuroanatomy of neurological disorders was an entirely different matter, involving a much more sophisticated conceptualization of notably the neocortex and its main outputs, notably the basal ganglia and the pyramidal motor system. It was said that psychiatry and neurology could be separated by the Sylvian fissure, and the two were generally believed to have little in common from a brain-oriented perspective. The funda-mental flaw in the scheme, however, was its conspicuous lack of concordance with many clinical observations. Obvious behavioral problems and psychiatric illnesses accompanying movement disorders (such as Parkinson's disease or Huntington's chorea), and the obviously abnormal movements in the clinical picture of psychiatric disorders across the spectrum from anxiety (tremor) to schizophrenia (tics, dystonias, dyskinesias, and catatonias) were a baffling conun-drum to clinicians. What the next generation of neuroanatomists has unraveled as introduced in this chapter (and described in more detail in Chapters 3–5) has revealed a totally different view of the forebrain and its connectivity. This con-ceptual change is shown in Table 1.

Thus, the realization that the whole of the cortical mantle projects to subcorti-cal basal ganglia structures allows for a closer integration of neurological and psychopathological thinking than in the past. The concept that the subcortical telencephalic nuclei are recipients of massive inputs from not only isocortex (neocortex), but also the nonisocortical (limbic lobe) parts of the cerebral cortex provides for a better way of understanding, in neuroanatomical terms, clinical phenomena from epilepsy to schizophrenia, and from laugh to cry.

CBOX 3 TABLE 1

Co-evolution of Neuroanatomical and Clinical Concepts

Old Version				
Allo (limbic) cortex	⟶	Hypothalamus	⟶	Psychiatry
Isocortex	⟶	Basal Ganglia	⟶	Neurology
New Version				
Allocortex	⟶	Basal Ganglia	⎤	
Isocortex	⟶	Basal Ganglia	⎦⟶	Neuropsychiatry

He added that the value of the limbic system "as a useful concept is correspondingly reduced," and it should therefore be abandoned. Instead, as opinions of what to include in the limbic system continued to proliferate, its boundaries became increasingly ambiguous. Curiously, the more the limbic system expanded, the more popular it became. For a scientist like Brodal, who was universally admired for his ability to illuminate the anatomy of the brain with precision and apt functional and clinical correlations, this must have seemed so very paradoxical. In short, the limbic system is a concept in perpetual search for a definition. Thus, there is little wonder why ambiguity seems to be among its most enduring qualities.

During the last 10 to 15 years, however, a few scientists have seriously addressed the shortcomings of the limbic system as a theoretical framework for emotional functions (Swanson, 1987; Kötter and Meyer, 1992; Isaacson, 1992; LeDoux, 1996; Blessing, 1997; Damasio, 1998). Based on a critical analysis of the limbic system, LeDoux (1996) has concluded that some of the major premises upon which MacLean developed his theories—for example, the triune brain concept, and the idea of hippocampus as a key player in emotional functions—are fundamentally wrong. LeDoux also has suggested that a global theory of emotion, such as embodied in the limbic system, does not promote an understanding of specific emotions. With reference to his own pioneering research on fear, LeDoux makes the point that different emotions are involved in different survival functions such as defense, appetite, reproduction, and so forth, and he suggests that different emotions involving partly different brain circuits need to be studied separately. Likewise, Damasio (1998), partly based on his studies of self-generated emotions (Damasio et al., 2000), also makes the point that different emotions are related to different neuronal circuits and adds that many structures outside the commonly accepted boundaries of the limbic system are involved in processing emotion and feeling. For example, the "greater limbic system" of Nieuwenhuys (1996), which was mentioned briefly in Chapter 1, contains structures—monoamine nuclei, periaqueductal gray, and other nuclei in the brainstem and spinal cord—that Damasio regards as lying outside the boundaries of what he considers to be the limbic system. In contrast, other structures of apparent importance for emotional functions, like orbitomedial prefrontal cortex and somatosensory cortices, are not included in the "greater limbic system" of Nieuwenhuys. These few examples illustrate the confusion that surrounds the limbic system concept even among pioneering neuroscientists. There are no generally accepted boundaries of the limbic system.

Nonetheless, others have continued to staunchly defend MacLean's triune brain theory and limbic system. Many of the relevant arguments are coherently summarized in *The Evolutionary Neuroethology of Paul MacLean* (Cory and Gardner, 2002), a book honoring MacLean's contributions in the field of neuroethology. For example, Jak Panksepp (2002), in his contribution to this book, suggests that those who criticize the limbic system concept have not made the effort to understand what MacLean's studies mean in the field of behavioral neuroscience. Panksepp's firmest support concerns MacLean's views on emotional feelings, a field in which Panksepp considers MacLean to be "a sure and guiding star." One is inclined to agree with Panksepp's statement that "... the main practical question about the limbic system concept concerns whether it *pointed* [our italics] effectively toward the main brain areas essential for emotionality." This seems fair enough; the limbic system, whatever version one prefers, does give a general idea of the parts of the brain involved in emotions. But Panksepp, being aware that a growing number of distinct neuronal circuits related to emotional functions already have been identified, goes beyond this by listing these "neural correlates" of specific emotions and partial functions in a continuously expanding "limbic neurogeography." One is inclined to rebut, however, by noting that many recent books in basic and clinical neuroscience with focus on emotional behavior and neuropsychiatric disorders (e.g., Rolls, 1999; McGinty, 1999; Charney et al., 1999; Zigmond et al., 1999; Shinnick-Gallagher et al., 2003; Swanson, 2003; Gazzaniga, 2004; Koob and Le Moal, 2006), not to mention research papers too numerous to list, deal effectively with these subjects without reference to an ambiguous limbic system concept. The reality of the situation can hardly escape anybody struggling to come to terms with the relevance of the limbic system in modern neuroscience. In conclusion, it is tempting to reiterate Brodal's wry warning that the increasingly inclusive nature of the limbic system in the end makes it a less useful guide to the analysis and study of emotional functions (Brodal, 1969). Indeed, the tide may soon turn in favor of Brodal's suggestion to abandon the limbic system concept, if it hasn't already.

2.3 NEW ANATOMICAL DISCOVERIES PROVIDE AN ALTERNATIVE TO THE LIMBIC SYSTEM

We will add our voice to the growing chorus questioning the usefulness of the limbic system as a guide in modern neuroscience research and psychiatry. We will argue that, beyond the conceptual inadequacy that leads to the almost perpetual expansion of the limbic system, the anatomical basis on which MacLean constructed his concept has been undermined by paradigm shifts resulting from experimental neuroanatomical work since the early 1970s. This work has altered our understanding of the fundamental organization of the forebrain. A couple of essential concepts that more or less presaged new thinking about forebrain organization are introduced in the following sections. Insofar as these constructs form the basis for a systems-based framework for the study of emotional functions and neuropsychiatric disorders, they will be the focus of more detailed discussion in the remaining chapters.

Ventral Striatopallidum

Investigations of the neuroanatomical organization of the basal forebrain in the 1970s showed that the ventral parts of the striatal complex, including the accumbens[2] and striatal territories in the olfactory tubercle, extend to the ventral surface of the brain and receive prominent allocortical and mesocortical (i.e., "limbic" cortical) inputs, just as the dorsal parts of the striatal complex receive massive projections from the neocortex (i.e., isocortex). Indeed, the ventral striatopallidal system, just like its dorsal counterpart, is characterized by massive, topographically organized cortico-striatopallidal-thalamocortical reentrant circuits (Heimer and Wilson, 1975). Although these new discoveries were duly reported by MacLean (1990), the fundamental principle of forebrain organization that emerges from them—that is, that the pattern of neural connections that characterizes basal ganglia also is expressed by "limbic" structures—is not reflected in MacLean's presentation of the limbic system in his book on the triune brain.

This is not to say, however, that MacLean and other proponents of the limbic system concept do not regard the accumbens as an important part of the limbic system. In historical context, the accumbens and olfactory tubercle, together referred to as the "olfactostriatum" by C. Judson Herrick (1926), were considered to serve as an intermediary between rostral forebrain areas and the medial forebrain bundle, a main conduit to the hypothalamus and brainstem tegmentum. Herrick stated with regard to the olfactostriatum that "its efferent fibers go in part to globus pallidus and thence to the cerebral peduncle, and in part they go out directly into the medial forebrain bundle" (Herrick, 1926). Later, Gordon Mogenson stuck with this theme in an oft-quoted article in which the accumbens was portrayed as a "limbic-motor" interface (Mogenson et al., 1980). In the concept of the ventral striatopallidal system, on the other hand, the accumbens is regarded not as a "limbic-motor" interface but simply as an integral component within a large system of cortico-basal ganglia-thalamocortical reentrant circuits that also includes the caudate nucleus, putamen, and olfactory tubercle. This will be discussed in greater detail in the next chapter. As intimated by Herrick, however, it is important to emphasize that ventral striatum, in contrast to dorsal striatum, does indeed give origin to descending projections to the lateral hypothalamus and midbrain tegmentum. These downstream connections are emblematic of the close relationship of one particular part of the ventral striatum—that is, the so-called

---

[2]The authors will freely use the term "accumbens" in the place of "nucleus accumbens" or "nucleus accumbens septi" throughout the book in deference to the demonstrated absence of detectable boundaries separating this structure from other parts of the striatal complex, including the caudate nucleus, putamen, and olfactory tubercle. This is discussed in greater detail in section 3.1.2 in Chapter 3.

shell of the accumbens, with another product of the histotechnical advances of the 1970s and 1980s—the extended amygdala—that we preview in the next section.

Before introducing the concept of extended amygdala, however, Mogenson's idea of a "limbic-motor" (i.e., "emotional-motor") interface requires some additional commentary. It is, perhaps, no coincidence that the word *emotion* is six-sevenths *motion*. Emotional expression invariably combines somatic motor, autonomic motor, and secretory motor—that is, endocrine—activity. Indeed, the close links between motion and emotion are very apparent in neuropsychiatric parlance—for example, in the use of the term *psychomotor* with respect to epilepsy or depression (as in psychomotor retardation). Many, if not all, motor disorders have psychiatric symptoms, and all psychiatric disorders reveal to the careful observer some accompanying motor component. If it is conceded that neural circuitry subserves both emotion and movement, a neural substrate for this apparent merging of neuropsychiatric and motor function should be discoverable, and, indeed, one favorite solution to the problem hearkens back to a "limbic-motor" interface. An expanded embodiment of such an interface (i.e., one that subsumes the accumbens limbic-motor interface described by Mogenson) refers to an "emotional motor system" (Holstege, 1992), which was envisioned as the principal effector apparatus of the limbic system and, by definition, is an integral part of the "greater" limbic system of Nieuwenhuys (1996). Like the limbic system concept, the "emotional motor system" (or, for that matter, the very idea of a specific "emotional-motor" interface) has a popular appeal. As we will discuss at some length in Chapters 3–5, however, neither the description of a special "limbic-motor" interface nor the identification of a separate "emotional motor system" adequately addresses this multifaceted problem.

Extended Amygdala

The discovery of the extended amygdala (de Olmos and Ingram, 1972) and its elaboration as a functional-anatomical entity (Alheid and Heimer, 1988) is another subject of profound importance for understanding the organization of the forebrain. On the basis of a special silver staining technique, de Olmos and Ingram (1972) identified a prominent neuroanatomically and connectionally distinct continuum that includes the centromedial nucleus of the amygdala and bed nucleus of the stria terminalis and also occupies a substantial stretch of basal forebrain territory extending between these two structures and thus interconnecting them. Since this concept establishes the bed nucleus of the stria terminalis as an integral part of the extended amygdala, it is important to note that MacLean included the bed nucleus of the stria terminalis in his concept of septum (see Fig. 1.2). This is unfortunate, since it ignores several distinct neuroanatomical differences between the septal nuclei and the bed nucleus of stria terminalis. Indeed, the bed nucleus has an abundance of characteristic cytoarchitectural, neurochemical, and connectional features that clearly distinguish it not only from the septal nuclear complex but also from adjoining parts of the ventral striatopallidal system (Alheid, 2003). These same features, however, so much resemble those observed in the centromedial nucleus of the amygdala that the bed nucleus seems best characterized as an extension of that structure—hence the name of the entire complex: extended amygdala. While the concept of the extended amygdala has yet to receive universal acceptance (Swanson, 2000; see also Basic Science Box 4), it has been amply confirmed in other laboratories (see next chapter), and its heuristic value is well established (McGinty, 1999; Koob and Le Moal, 2006).

2.4 CONCLUSION

It is apparent that limbic system proponents and opponents alike are looking for something beyond the limbic system. One of LeDoux's major complaints is that MacLean "packaged . . . the entire emotional brain and its evolutionary history into one system." LeDoux maintains that different survival functions must be subserved by different emotional systems that need to be studied separately. To this end, the idea of a limbic system is of little use. A similar idea expressed by some supporters of the limbic system (e.g., Morgane et al., 2005) is that "the anatomy of partial functions need to be better pieced together than they have in the past." In this chapter, we have highlighted some reasons for abandoning the limbic system concept and suggest that MacLean's anatomical basis for the limbic system has been significantly undermined by discoveries in the comparative and experimental neuroanatomy of the forebrain, which were fueled by histotechnical advances in the 1960s and 1970s. These advances gave birth to the concepts of the ventral striatopallidal system and extended amygdala, which have changed the way we look at the functional-anatomical organization of the basal forebrain. These discoveries are not easily reconciled with any of the current models of the limbic system and are described in further detail in Chapter 3. We hope that this emerging appreciation of the anatomy of the basal forebrain together with recent modifications in our thinking about Broca's limbic lobe (Chapter 4) will facilitate a transition from reliance on a limbic system to a commitment to a systems-oriented approach to the study of emotional functions and adaptive behaviors (Chapter 5).

3

THE ANATOMY OF THE
BASAL FOREBRAIN

3.1 INTRODUCTION

As an introduction to a review of the systems anatomy of the basal forebrain, we will briefly focus attention on four anatomical structures: the septum, accumbens, substantia innominata, and amygdala (amygdaloid body). Besides orbitofrontal and temporal cortical regions, including the superficially obvious continuations of the olfactory system, these four structures, elusive or poorly demarcated as they may be, have for many years defined the telencephalic part of the basal forebrain for most neuroscientists. The septum, accumbens, and amygdala are key structures of the limbic system. Whereas interest in the septum has perhaps not kept pace, the accumbens and amygdala have continued to figure prominently in the neuroscience literature, ranking among the most popular brain structures in studies of emotion, motivation, substance abuse, and neuropsychiatric disorders. A pictorial survey of the human basal forebrain is presented in Fig. 3.1. These highly schematic diagrams reflect our current understanding of the anatomical organization of the basal forebrain to be reviewed in the main part of this chapter, but they will also serve a useful purpose in this introductory survey.

3.1.1 Septum: A Structure with Many Faces

Reference to the septum is likely to evoke different images, depending on one's viewpoint. Clinicians will likely think of the septum pellucidum, a thin sheet of nervous tissue (in reality, two, often with a cavum septi pellucidi between them) separating the anterior horns of the two lateral ventricles in the human brain. Basic scientists familiar with experimental animals like the rat will no doubt recollect several distinct nuclear groups housed within plump, paired column-like bodies, that, like the septum pellucidum in the human, stand as a partition between the anterior parts of the lateral ventricles. Although the appearance of the septum seems dramatically different in the human and rat, the difference is more superficial than real. In reality, the systems anatomy of the septal region is quite similar in the two species. Whereas the septum pellucidum in the human contains mostly fibers and glia cells, collections of septal neurons in the human that correspond to the rat septal nuclei are located primarily in the paraterminal gyrus, which is directly continuous with the ventral part of the septum pellucidum in front of the anterior commissure (see Fig. 4.4 in Chapter 4). Therefore, the septal nuclei in the human are often referred to as the precommissural septum.

But this is only part of the problem with the septum, another difficulty being that investigators over the years have tended to lump it, or the so-called "septal area," with a number of other structures. MacLean (1990), for instance, included the bed nucleus of the stria terminalis (Figs. 3.1B and C) in his concept of septum (p. 288). The confusion surrounding this structure becomes even more understandable when one considers the writings of Heath (1954), who, in his studies of psychiatric patients, described a "septal region" close to the midline and extending from the rostral tip of the anterior horn of the lateral ventricle to the level of the anterior commissure (roughly 2 cm in the rostrocaudal dimension). This volume of neural tissue includes several anatomically distinct basal forebrain structures—for example, part of the corpus callosum and subcallosal cortex (a part of the limbic lobe; see Chapter 4), a medial part of the ventral striatum (Fig. 3.1A), the diagonal band of Broca, septal nuclei and, in all likelihood, also parts of medial preoptic area and bed nucleus of stria terminalis.

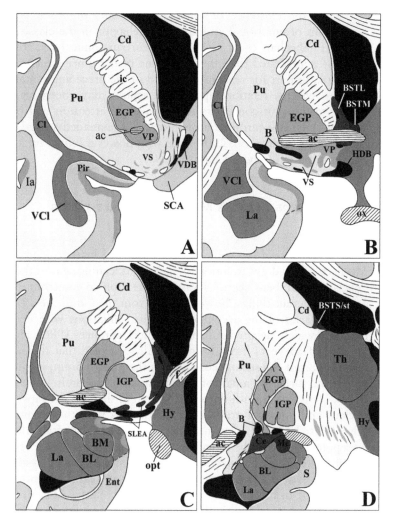

FIGURE 3.1 Schematic drawings showing the human basal forebrain in a series of coronal sections starting rostrally at the level of the accumbens in A and ending at the level of the caudal amygdala in D. Color coding: striatum—light blue; pallidum—salmon; extended amygdala central division—yellow; extended amygdala medial divsion—green; basal forebrain magnocellular complex—brown; primary olfactory areas—pink; Abbreviations: ac—anterior commissure; B—basal nucleus of Meynert; BL—basolateral amygdala; BM—basomedial amygdala; BSTL and BSTM—lateral and medial divisions of the bed nucleus of the stria terminalis; BSTS/st—supracapsular/stria terminalis division of the extended amygdala (shown in yellow); Cd—caudate nucleus; Ce—central nucleus of the amygdala; Cl—claustrum; EGP—external segment of the globus pallidus; Ent—entorhinal cortex; f—fornix; Hy—hypothalamus; IGP—internal segment of the globus pallidus; La—lateral amygdala; Me—medial nucleus of the amygdala; opt—optic tract; ox—optic chiasm; Pir—piriform (primary olfactory) cortex; Pu—putamen; S—subiculum; SCA—subcallosal area; SLEA—sublenticular extended amygdala (shown in yellow [central division] and green [medial division]); Th—thalamus; VCl—ventral claustrum; VDB—vertical limb of the diagonal band of Broca; VP—ventral pallidum; VS—ventral striatum. (Art by Medical and Scientific Illustration, Crozet, Virginia; reprinted from Heimer et al., 1999 with permission.) See color plate.

Despite the existence of abundant accounts of septal contributions to expressions of emotional and adaptive behavior in the classical and modern literature (Sheehan et al., 2004), its neuroanatomical relationships are less precisely appreciated than those of some of the other structures discussed in this book, in part for reasons discussed in the preceding sentences. Nonetheless, the systems relationships of the septum have been regarded as quite similar to those of constructs to which more attention will be given in our account—that is, ventral striatopallidum and extended amygdala (Alheid and Heimer, 1988; Swanson, 2000). We will, however, come back to septal relationships, particularly in Chapter 5.

3.1.2 Nucleus Accumbens Septi: A Magical Structure with a Misleading Name

More than a century ago, the Austrian psychiatrist-neurologist Theodor Meynert (1872) drew attention to "a nucleus leaning against the septum"—the "nucleus accumbens septi." Drawing attention to something and defining it are two different things, however, and from then until the 1970s and 1980s, the boundaries of the accumbens and its affiliations with nearby structures had been widely debated (Chronister and DeFrance, 1980). However, during the last 20–25 years it has been generally accepted that the accumbens is an integral part of the striatal complex. In the human brain, it is located close to the ventral surface in a part of the rostral basal forebrain known as the fundus striati. Here, underneath the anterior limb of the internal capsule, the caudate nucleus and putamen are directly continuous with the accumbens and each other (Figs. 3.1A and 3.2), and it is not possible to precisely identify boundaries separating any of these structures from each other. Thus, as will be evident in our more detailed review of the ventral striatopallidal system (see following), the accumbens is not a separate nucleus and cannot be precisely delimited from the rest of the striatal complex. To the contrary, it is an integral part of the striatal complex, within the so-called ventral striatum, which, incidentally, also includes neighboring ventral parts of the caudate nucleus and putamen.

The accumbens contains an intriguing and functionally important divide between a centrally located core and a shell that surrounds this core on its medial, ventral, and ventrolateral aspects. The core-shell dichotomy has attracted a lot of attention, especially among basic scientists interested in emotional functions and adaptive behaviors and will be discussed in some detail later. For the moment, suffice it to say that the important dividing line in this part of the ventral striatum is apparently located between the core and shell rather than between the accumbens and the rest of the striatal complex.

The accumbens owes much of its fame, particularly in the field of psychiatry, to the work of Arvid Carlsson and his colleagues (1958). The discovery

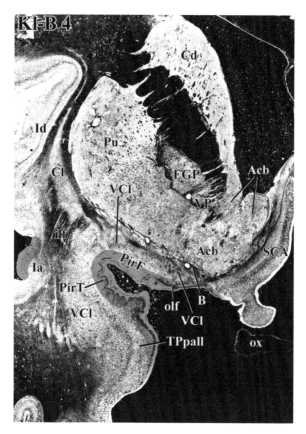

FIGURE 3.2 Klüver-Barrera stained frontal section through the human brain at the level of the optic chiasm. Primary olfactory cortex is highlighted in pink. Abbreviations: Acb—accumbens; B—basal nucleus of Meynert; Cd—caudate nucleus; Cl—claustrum; EGP—external segment of the globus pallidus; Id—dorsal insular cortex; Ia—anterior insular cortex; ilf—inferior longitudinal fasciculus; olf—olfactory tract; ox—optic chiasm; PirF—piriform cortex, frontal portion; PirT—piriform cortex, temporal portion; Pu—putamen; SCA—subcallosal area; TPpall—temporopolar periallocortex; VCl—ventral claustrum; VP—ventral pallidum. (Art by Medical and Scientific Illustration, Crozet, Virginia; reprinted from Sakamoto et al., 1999 with permission.) See color plate.

of dopamine as a neurotransmitter and the proposal that chlorpromazine and haloperidol exert their influence by blocking dopamine receptors (Carlsson and Lindquist, 1963) provided the basis for the dopamine hypothesis of schizophrenia. The suggestion that schizophrenia might be due at least in part to disturbances in the so-called mesolimbic dopaminergic (DA) system, comprising a massive system of dopaminergic projections from the ventral mesencephalon to the frontal cortex and a number of forebrain structures, most notably the accumbens (Matthysse, 1973; Stevens, 1973), propelled the accumbens to become one of the most studied "limbic" structures in the context of neuropsychiatric disorders.

Another likely basis for the popularity of the accumbens is the suggestion that it serves as a "limbic-motor" interface (Mogenson et al., 1980), where "limbic" processes supposedly gain access to the motor system. Persuasive evidence in support of this theory came from tract-tracing studies in the rat brain, which indicated that the accumbens projects to the globus pallidus (Swanson and Cowan, 1975). Globus pallidus, with prominent projections to the ventral anterior–ventral lateral (VA–VL) "motor" nucleus of the thalamus (and hence to the motor cortex), as well as to brainstem somatomotor centers, is appreciated foremost for its relation to the motor system. Since the accumbens was known to receive major inputs both from the amygdala and hippocampus, which are key structures in anybody's definition of the limbic system, and since locomotor activation accompanies a variety of pharmacological manipulations of the accumbens, Mogenson concluded accordingly. The popularity of his hypothesis seems to have reinforced the impression that the accumbens is independently and seemingly magically capable of transforming emotion to motion.

Further elaboration of the extrinsic connections of the ventral striatopallidal system suggested that it is misleading to designate the accumbens as a specialized limbic-motor interface based on a projection leading to the thalamic VA–VL "motor" nuclei via the globus pallidus, or at least this is a diversion from a more robust and important connectional relationship. Numerous studies now have shown that the ventral pallidum projects much more robustly to the mediodorsal nucleus of the thalamus, which is closely associated by direct connections more with the prefrontal rather than the premotor cortex. This discovery paved the way for our current appreciation of parallel frontal cortico-subcortical reentrant circuits, which are now promoted as a basic framework for explaining the symptomatology of neuropsychiatric disorders (Mega and Cummings, 1994; Lichter and Cummings, 2001).

In this process of discovery, the accumbens has lost its status as an independent functional-anatomical unit, not to mention its standing as a nucleus. The notion of accumbens as a critical "limbic-motor" or "emotional-motor" interface, furthermore, has become equally dubious. Not only ventral striatopallidum, but also septum, amygdala, preoptic region, hypothalamus, and some cortical regions in the limbic lobe (Chapter 4)—practically all major forebrain regions—are involved in emotional, motivational, and motor activating functions because all have more or less direct relationships with motor effector structures in the brainstem. This situation gave rise to the idea of an "emotional motor system" (Holstege, 1992), which was touched on in Chapters 1 and 2 and will be discussed further in Chapter 5.

3.1.3 Substantia Innominata: The Neurologist's Equivalent of the Cartographer's "Terra Incognita"

As already mentioned, the accumbens cannot be precisely delineated from any of the surrounding parts of the striatal complex; it is an integral part of a larger ventral striatal territory. Much of ventral striatum is located underneath the temporal limb of the anterior commissure (Figs. 3.1B and 3.3), invading, as it were, an obscure part of the brain that, without apparent internal or external borders, has for close to 200 years been recognized by the name of substantia innominata.[1] The most conspicuous component of the substantia innominata, or "basalis region," is the basal nucleus of Meynert (colored red in Fig. 3.1), which stands out because it houses an aggregation of large hyperchromatic cholinergic cells (Fig. 3.4). But the basal nucleus of Meynert, sometimes referred to as the nucleus of the substantia innominata, is only one of several neuronal components in this area, and efforts to elucidate more fully the anatomical organization of the substantia innominata, let alone its functional organization, had until recently met with considerable difficulties. Therefore, the substantia innominata, which also includes the basal, subpallidal region (between the hypothalamus and the amygdaloid body in Fig. 3.1C), has remained one of the most mysterious regions in the human brain in spite of the fact that it has often been the focus of attention in the neuropsychiatric literature (Heimer et al., 1997). Recent histotechnical advances, however, finally opened up this region to fruitful anatomical explorations, and the term *substantia innominata* is gradually disappearing, as its main parts have been identified and shown to be aligned not only with a widely dispersed, more or less continuous collection of cholinergic and noncholinergic neurons referred to as the magnocellular basal forebrain complex (which includes the basal nucleus of Meynert) but also with extensions of the basal ganglia and amygdaloid complex (Alheid and Heimer, 1988).

3.1.4 Amygdala: What Is It?

This is a pertinent question (Swanson and Petrovich, 1998), and it may come as a surprise that we still debate the anatomical organization of such a

---

[1]Substantia innominata is often referred to as the "substantia innominata of Reichert" for the simple reason that in his atlas of the human brain, Reichert (1859) left part of the area underneath the temporal limb of the anterior commissure unnamed. It may be more appropriate, however, to attribute the term to the German anatomist Johann Christian Reil (1809), who included the region in what he referred to as "die ungennante Mark-Substanz," which he observed by stripping the optic tract caudally toward the region of the lateral geniculate body. Reil wanted to postpone the naming of the region until he had a better idea of its functional organization.

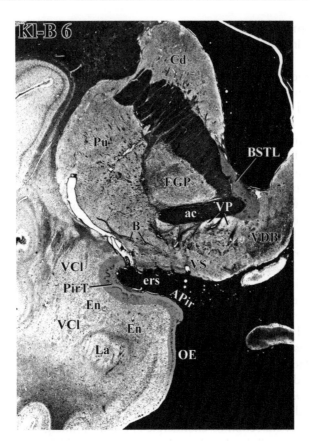

FIGURE 3.3 Klüver-Barrera stained frontal section through the human brain at a level just rostral to the crossing of the anterior commissure. Primary olfactory cortex is highlighted in pink, central division of the extended amygdala in yellow. Abbreviations: ac—anterior commissure; Acb—accumbens; APir—amygdalopiriform transition cortex; B—basal nucleus of Meynert; BSTL—lateral division of the bed nucleus of the stria terminalis; Cd—caudate nucleus; Cl—claustrum; EGP—external segment of the globus pallidus; En—endopiriform nucleus; ers—endorhinal sulcus; LA—lateral amygdala; PirT—piriform cortex, temoral portion; Pu—putamen; TPpall—temporopolar periallocortex; VCl—ventral claustrum; VDB—vertical limb of the diagonal band of Broca; VP—ventral pallidum; VS—ventral striatum. (Art by Medical and Scientific Illustration, Crozet, Virginia; reprinted from Sakamoto et al., 1999 with permission.) See color plate.

well-known entity as the amygdala. However, despite the clarity of the classical descriptions of the amygdala complex and the protestations of a few neuroscientists who read the older literature, the amygdala over the years has frequently, seemingly intractably, been portrayed in publications as a more or less homogeneous functional-anatomical unit. Effort to apply what one might call an anatomical systems analysis to the amygdala, when considering its functional and clinical significance, has been limited, to say the least.

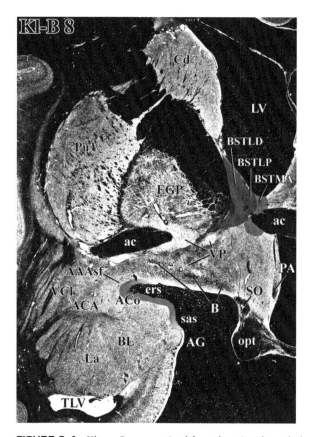

FIGURE 3.4 Klüver-Barrera stained frontal section through the human brain at the level of the crossing of the anterior commissure. Olfactory bulb projections to the cortical amygdaloid nucleus are highlighted in pink, central and medial divisions of the extended amygdala in yellow and green, respectively. The large hyperchromatic cells of the basal nucleus of Meynert (B) form particularly large conglomerates at this level. Abbreviations: AAAsf—anterior amygdaloid area, superficial division; ac—anterior commissure; ACA—amygdaloclaustral area; Acb—accumbens; Aco—cortical nucleus of the amygdala; AG—ambiens gyrus; APir—amygdalopiriform transition cortex; B—basal nucleus of Meynert; BL—basolateral thalamic nucleus; BSTMA, BSTLD, and BSTLP—medial division, anterior part and lateral division, dorsal and posterior parts of the bed nucleus of the stria terminalis; Cd—caudate nucleus; ers—endorhinal sulcus; LA—lateral amygdala; LV—lateral ventricle; opt—optic tract; Pu—putamen; SO—supraoptic nucleus; TLV—temporal horn of the lateral ventricle; Pa—hypothalamic paraventricular nucleus; sas—semiannular sulcus; TPpall—temporopolar periallocortex; VCl—ventral claustrum; VDB—vertical limb of the diagonal band of Broca; VP—ventral pallidum; VS—ventral striatum. (Art by Medical and Scientific Illustration, Crozet, Virginia; reprinted from Sakamoto et al., 1999 with permission.) See color plate.

The amygdala is located in the anterior part of the temporal lobe in front and above the temporal horn of the lateral ventricle, where it abuts the rostral part of the hippocampus, a beautifully curved body of allocortex that forms the floor and medial wall of the temporal horn. It is a more or less almond-shaped structure when cut either in the coronal or horizontal plane during macroscopic brain dissections—hence the term *amygdala*: "almond nucleus," introduced by Burdach (1819) almost 200 years ago. It is important to emphasize that Burdach meant *amygdala* to refer only to what we now know as the basolateral amygdaloid complex (McDonald, 2003), which in the human is by far the largest part of the amygdala (Figs. 3.1C and D). In contrast, Brockhaus (1938) subdivided the amygdala into an "amydaleum proprium"—that is, the amygdaloid body in the strict sense (Burdach's amygdala, consisting of the laterobasal-cortical amygdala)—and a "supra-amygdaleum" (centromedial amygdala and anterior amygdaloid area), which according to several classical anatomists extends without clear demarcation into the subpallidal part of the basalis region. Based on comparative-anatomical and developmental studies, J. B. Johnston (1923) also divided the amygdala into a centromedial part and a laterobasal-cortical complex. Johnston, however, realized that centromedial amygdaloid nucleus extends medialward through the substantia innominata to establish continuity with the bed nucleus of the stria terminalis (Figs. 3.1C and D) and, furthermore, pointed out that this centromedial amygdala-bed nucleus complex has a developmental origin distinct from that of the laterobasal-cortical complex.

Numerous students of the amygdala have observed that, in terms of neuronal structure and neurochemical composition, the structures comprising the laterobasal-cortical nucleus complex closely resemble the cerebral cortex. In contrast, the centromedial nucleus is striatopallidal-like in its cellular and neurochemical composition and connections. This basic subdivision of the amygdala into a large, "cortical-like" laterobasal-cortical part and a subcortical centromedial part that is continuous with the bed nucleus of stria terminalis is consistent with the classical descriptions of Burdach and Johnston and represents the starting point for the concept of the extended amygdala, which is an important component of our current perspective on the basal forebrain and fundamental to our definition of the limbic lobe (see also Chapter 4).

3.1.5 Where Do We Go from Here?

During the last quarter-century, a new systems anatomy has replaced the substantia innominata, nucleus accumbens, and amygdala, not to mention the limbic system, as a framework for studies of physiological and behavioral functions typically affected in neuropsychiatric disorders. One of our goals is to focus attention on these recent anatomical findings that better illuminate the fundamental organization of the forebrain and its neuronal circuits. Thus, in the following sections of this chapter, we will discuss three of the major functional-anatomical macrosystems of the basal forebrain: the ventral striatopallidal system, the extended amygdala, and the magnocellular basal forebrain system (including the basal nucleus of Meynert). For some perspective, it may be useful to recall that it is only during the last 25 years—almost 200 years after its introduction into the anatomical literature—that the term *substantia innominata* began to disappear from the literature as its various parts were shown to belong to nearby and better-defined anatomical systems (Alheid and Heimer, 1988).

3.2 THE VENTRAL STRIATOPALLIDAL SYSTEM

3.2.1 A Paradigm Shift Involving the Higher-Order Olfactory Connections

Since the crucial experiments that led to the concept of the ventral striatopallidal system were performed in the rat brain, we will introduce the ventral striatopallidal system with a few illustrations of histological preparations from the rat brain. Figure 3.5A depicts the "limbic system" paradigm of forebrain organization. In accordance with the limbic system concept, projections from "limbic" allocortical areas, including the olfactory cortex, hippocampus formation, and the cortical-like basolateral amygdala (broken arrow), had long been believed to terminate primarily in the medial forebrain bundle area in the lateral hypothalamus. On the other hand, the isocortex (a preferred term for neocortex; see footnote 1, Chapter 2) was known to project to the basal ganglia. Anatomical tract-tracing experiments (Heimer, 1972) reviewed in Basic Science Box 2, however, revealed a completely different situation (Fig. 3.5B). Both the cortical-like laterobasal amygdaloid complex and allocortex, represented in the drawing by the olfactory cortex, project massively not to the region of the medial forebrain bundle in the lateral preoptic-lateral hypothalamic area but to prominent extrahypothalamic ventral extensions of the basal ganglia—that is, to ventral striatum. Ventral striatum, in turn, gives rise to a massive striatopallidal projection to ventral pallidum (Heimer and Wilson, 1975).

An important outcome of the research described in the Basic Science Box 2 was the identification of the olfactory tubercle as a basal ganglia structure. That is, the olfactory cortex gives rise to a corticostriatal projection to the superficially located striatal parts of the tubercle, which, in turn, have massive striatopallidal projections to pallidal territories in deep parts of the tubercle and the subcommissural region. Now, to suddenly focus in this account on the olfactory tubercle and its connections, rather than the accumbens, may seem confusing, especially since the accumbens, as mentioned earlier, is the

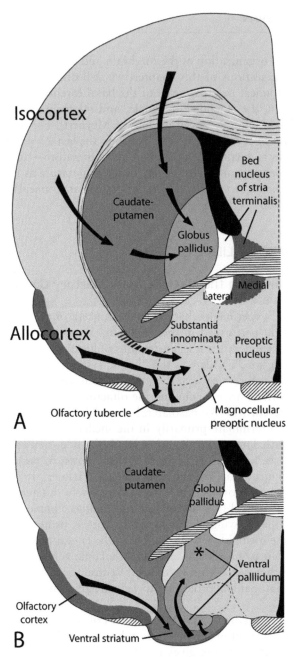

FIGURE 3.5 Schematic drawings illustrating in A the "limbic versus basal ganglia dichotomy," which has been discredited following the discovery of the ventral striatopallidal relations of the olfactory system shown in B. Projections from limbic cortical area, including the lateral-basal complex of the amygdala (broken arrow in A) were previously believed to teminate in the medial forebain bundle area in the region of the magnocellular preoptic nucleus, whereas the isocortex was known to project to the basal ganglia (caudate-putamen and globus pallidus). The image in B reflects the olfactory tubercle as a ventral extension of the striatal complex—that is, the ventral striatum, which in turn projects to a ventral extension of the globus pallidus, or the ventral pallidum. The asterisk in B indicates the area of terminations for the accumbens projection at this level. (Graphic art by Medical and Scientific Illustration, Crozet, VA; reprinted with permission from Heimer, 2003.)

BASIC SCIENCE BOX 2

Discovery of the Ventral Striatopallidal System

The ventral striatopallidal system, which signifies a major ventral expansion of the basal ganglia to the ventral surface of the mammalian brain, was discovered in the process of tracing olfactory connections in the rat. The rat is a macrosmatic animal—that is, it has a well-developed sense of smell—which is reflected in part by the fact that the entire basal surface of the rat brain (the areas in Fig. A) receives direct input from the olfactory bulb. The olfactory tubercle, in the figure, was generally considered to be a modified part of the olfactory cortex, and in accordance with the limbic system concept, the understanding was that projections from the olfactory cortex and cortical-like olfactory tubercle (see "deep olfactory radiation" in Fig. A) converged massively on the anterolateral hypothalamus (magnocellular preoptic nucleus in Fig. 3.5A). However, by tracing the projections from some of these olfactory areas with the aid of a method that stains degenerating axons following an experimental lesion in the area of origin of the axons, we discovered something that turned the "limbic system–basal ganglia" dichotomy on its head. Following a lesion in the olfactory cortex, we found that hardly any fibers degenerated in the anterolateral hypothalamus; rather, degenerating fibers projected massively to surrounding extrahypothalamic regions (Fig. B), including the olfactory tubercle, ventral pocket of the caudate-putamen, and cell bridges between these two structures (Heimer, 1972).

The next question presented itself automatically: What happens if we make a superficial lesion in the olfactory tubercle? Figure C shows the result following a laminar heat lesion of the

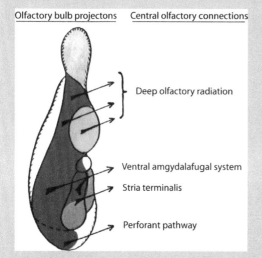

Olfactory bulb projectons   Central olfactory connections

Deep olfactory radiation

Ventral amgydalafugal system

Stria terminalis

Perforant pathway

BSBOX 2 FIGURE A Diagram of the basal surface of the rat brain illustrating the areas that receive direct projections from the olfactory bulb (bulbous top, light gray) and the outputs from those areas (arrows) as they were understood to be organized about 1970. Recipient areas of the main olfactory bulb are shown in dark and light gray and of the accessory olfactory bulb in intermediate gray. It was generally accepted that the olfactory tubercle (light gray oval) was a modified part of the olfactory cortex and that the deep olfactory radiation converges on the anterolateral hypothalamus. This understanding turned out to be incorrect (see BSB 2 Fig. B).

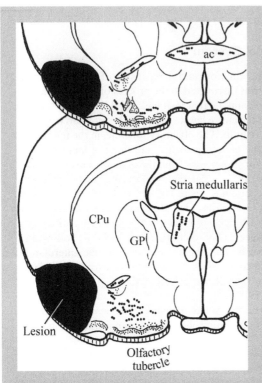

BSBOX 2 FIGURE B A critical experiment showing degeneration in the rat basal forebrain following the creation of a lesion in the olfactory cortex 2 days earlier. The small and large dots indicate terminal degeneration and degenerating fibers of passage, respectively. The lesion is shown in black. The medium-sized cellular regions of the olfactory tubercle and cell bridges between the tubercle and subcommisural striatal pocket are riddled with terminal degeneration, whereas the area of the magnocellular preoptic nucleus contains only fibers of passage. Abbreviations: ac—anterior commissure; CPu—caudate-putamen. (Reprinted from Heimer, 2003, with permission.)

olfactory tubercle, produced by inserting a heated (70°C) silver plate beneath the tubercle from the lateral side. The reasoning behind the heated silver plate was that the heat would be just enough to destroy superficially located neurons in large numbers but leave the blood supply to deeper areas intact in order not to create deep lying lesions in regions of interest. As a result of this lesion, there was massive but well-contained terminal degeneration within what initially seemed to be part of the lateral hypothalamus. When we looked a little closer, however, using plastic-embedded sections examined with light and electron microscopy (Fig. D), we realized that the region of dense terminal degeneration belongs to a ventral extension of globus pallidus (Fig. 3.5B). Some of the most convincing evidence was provided by the ultrastructural analysis, which demonstrated that the region containing terminal degeneration was in every major aspect comparable to that of globus pallidus (Heimer and Wilson, 1975). The boutons that degenerated following a lesion in the tubercle, furthermore, were of the same morphological type as boutons that degenerate in the dorsal part of the pallidal complex after a lesion is made in the caudate-putamen. It became apparent that the olfactory tubercle is a striatal rather than a cortical structure, despite its laminar organization. We later realized, based on the results obtained by the aid of anatomical and histochemical methods (Millhouse and Heimer, 1984; Switzer et al., 1982), that even the pallidal complex, like the striatum, extends ventrally to include part of the olfactory tubercle, as indicated in Fig. 3.5B. In other words, the olfactory tubercle, which had long been considered to be a slightly modified medial continuation of the olfactory cortex, is instead an integral part of the basal ganglia (Heimer et al., 1995). The absence of projections from the tubercle to the laterally adjacent parts of the olfactory cortex (Haberly and Price, 1978) was consistent with this conclusion. Reciprocal association connections are a fundamental feature of cortical organization, and projections from the tubercle to the olfactory cortex would in all likelihood exist if the tubercle is an integral part of the olfactory cortex.

Note on the Choice of Method

Breakthroughs in neurobiology are often the result of histotechnical advances, and the discovery of the ventral striatopallidal system is no exception. The research just described was aided by silver methods developed in the late 1960s (see Basic Science Box 1 in Chapter 2). But there is a twist to this story. In the early 1970s, when these studies took place, a number of new methods were introduced, based on the biological principle of axonal transport following the injection

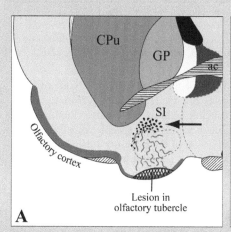
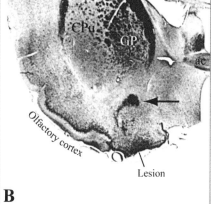

BSBOX 2 FIGURE C A. Diagrammatic representation of the characteristic pattern of degeneration in the general region of the medial forebrain bundle following the creation of a laminar heat lesion in a rat olfactory tubercle two days earlier. B. A matching histological preparation showing the lesion and area of terminal degeneration. Panel A shows that the degeneration is represented by terminal degeneration (arrow) in the region of the substantia innominata and by fibers of passage (wavy lines) in the area of the magnocellular preoptic nucleus. The matching histological section (B) is modified from Heimer and Wilson (1975). The terminal degeneration indicated by the arrow is shown at higher magnification in BSB Fig. C. Abbreviations: ac—anterior commissure; CPu—caudate-putamen; GP—globus pallidus; SI—substantia innominata. (Reprinted from Heimer, 2003, with permission.)

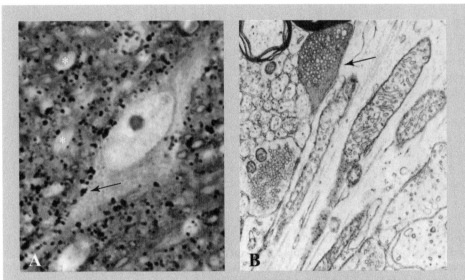

BSBOX 2 FIGURE D High-magnification light (A) and electron (B) micrographs depicting terminal degeneration in plastic-embedded sections of the rat brain. The section in A was stained with the Fink-Heimer technique two days following the creation of a lesion similar to the one shown in BSB 2 Fig C. It shows terminal degeneration at the site marked with an arrow in both panels in BSB 2 Fig C. The section in B illustrates the same site, also from a similarly lesioned brain, processed for electron microscopy. The structure indicated by the arrow is a striatopallidal bouton showing increased electron density and swollen vesicles indicative of terminal degeneration. (Reprinted from Heimer, 2003, with permission.) See color plate.

of a retrograde or anterograde tracer (e.g., HRP, PHA-L, or fluorescent tracer). None of these tracer methods, however, would have been able to display the striatopallidal link from the olfactory tubercle to the ventral pallidum in such a convincing fashion as the silver method. The great advantage of using an experimental silver method in this specific case is related to the fact that we could make an extensive superficial lesion of the olfactory tubercle (by using the heated silver plate as discussed earlier) without having to deal with the "fibers-of-passage" problem—that is, the destruction of fibers passing through the lesion from other regions of the brain. The use of the silver plate, furthermore, resulted in an extensive lesion, capable of producing massive terminal degeneration in a well-defined area (Fig. C). This was an important prerequisite for the subsequent ultrastructural analysis of synaptic relations, especially since an adequate preliminary mapping of degeneration in the light microscope was necessary in order to select the proper tissue fragment to be viewed using electron microscopy. The strategy of combining light- and electron-microscopic examination of degenerating synaptic fields on one and the same plastic-embedded specimen (Heimer, 1970) was key to the success of this study. When the overall pattern of the striatopallidal projections from the olfactory tubercle had been successfully displayed, we used a combination of recently developed tracer and immunohistochemical methods to elucidate additional details of this projection, including its participation in the cortico-subcortical reentrant circuitry (Young et al., 1984; Heimer et al., 1987; Zahm and Heimer, 1987; Zahm et al., 1987).

more popular ventral striatal structure in the context of functional-anatomical and neuropsychiatric correlations. However, the olfactory tubercle had traditionally been regarded as modified olfactory cortex projecting to the antero-lateral hypothalamus, and, as explained in Basic Science Box 2, the impetus critical to the conceptualization of a ventral striatopallidal system, of which accumbens is an important part, actually came from the combined light and electron microscopic analysis that served to disabuse us of misconceptions regarding the nature of the olfactory connections in the olfactory tubercle. That the accumbens projects to a pallidal region did not come as a big surprise (Heimer and Wilson, 1975; Swanson and Cowan, 1975), since the striatal nature of the accumbens had long been suggested (Fox, 1943) and cortico-striatal-like cortical input to the accumbens from the hippocampus had been confirmed repeatedly in neuroanatomical studies. Nonetheless, the striatal nature of the accumbens had been obscured in the shadow of the more widely accepted notion that it belongs to the "limbic system." It was the discovery that the olfactory tubercle of the rat in actuality comprises largely striatopal-lidum that advanced the opportunity to actualize "basal ganglia character" (i.e., striatopallidum) as the defining property of the accumbens. The asterisk in the salmon-colored depiction of ventral pallidum in Fig. 3.5B indicates the projection area of fibers from the accumbens.

The conceptual importance of the ventral striatopallidal system is not only as an expansion of the basal ganglia but, more significantly, as an expansion of the cortical output apparatus. What these findings meant is that the entire cortical mantle, not just the neocortex, projects to basal ganglia. The discovery that the allocortical (i.e., 3-layered) olfactory cortex and hippocampus formation project to the basal ganglia can to some extent be viewed as the final "missing link" in the establishment of a fundamental principle of forebrain organization—that all of cortex utilizes basal ganglia mechanisms. It may well be—indeed, seems likely—that proper cortical function *requires* access to basal ganglia mechanisms. That this principle was revealed by studies of olfactory pathways in the rat does not diminish its relevance for the overall organization of the mammalian, including human, brain. To the contrary, the study of a macrosmatic mammal—that is, with a well-developed and relatively accessible olfactory cortex—was in all likelihood the only realistic experimental platform with which this fundamental discovery could have been made.

Incidentally, the paradigm shift reflected in the new understanding of the striatopallidal nature of the olfactory tubercle was acknowledged early on in a popular paper entitled "An Anatomy of Schizophrenia" (Stevens, 1973). The fact that the olfactory cortex (often referred to as paleocortex) projects massively to the olfactory tubercle and ventral parts of caudate-putamen rather than the anterolateral hypothalamus (Heimer, 1972) was expressed by Stevens in a colorful metaphor involving two ornate Japanese fans (Fig. 6 in Stevens,

1973). The projection of "neocortical structures on caudate-putamen" was represented by a dorsal fan-like confluence. A similar ventral fan-like confluence represented projections from allocortical structures to ventral striatum (which Stevens referred to as "limbic" striatum), including "projection of amygdala to bed nucleus of stria terminalis, hippocampus to nucleus accumbens, and piriform cortex to olfactory tubercle."

3.2.2 The Ventral Striatum in the Human

Although experimental studies in the rat were instrumental to the discovery of the ventral striatum and ventral pallidum, it is important to emphasize that additional studies in other mammals including primates (Alheid and Heimer, 1988; Martin et al., 1991; Heimer et al., 1999; Haber and Johnson Gdowsky, 2004) have contributed significantly to our understanding of the ventral striatopallidal system in the human brain. The diagrams in Fig. 3.1 depict a series of coronal sections through the human forebrain from the level of the rostral part of ventral striatum (i.e., the accumbens, Fig. 3.1A) to the level of the caudal amygdala (Fig. 3.1D). Just in front of the crossing of the anterior commissure (Fig. 3.1A), the caudate nucleus and putamen are continuous with each other underneath the anterior part of the internal capsule. This broad continuity of striatal structures, which includes the accumbens (Fig. 3.2), is sometimes referred to as the fundus striati. Although the term fundus striati is not part of the official anatomical nomenclature, it actually reflects the existing situation better than "accumbens," which brings to mind a circumscribed structure (which the accumbens is not) that can be easily identified and separated from the rest of the striatum (which the accumbens can not). As discussed at the beginning of this chapter and indicated in Figs. 3.1A and 3.2, it is not possible to precisely delineate boundaries between the accumbens and other parts of the striatum. Accumbens is an integral part of the striatal complex and, more specifically, of what we now refer to as ventral striatum.

Structures lateral to the hypothalamus, beneath the temporal limb of the anterior commissure (Fig. 3.1B) and globus pallidus (Fig. 3.1C) and in the dorsal aspect of the amygdala within the temporal lobe (Fig. 3.1D), had long been regarded as adequately amorphous to be lumped together as the substantia innominata or basalis region. This dogma began to dissipate, however, when the rostral part of the basalis region shown in Figs. 3.3B and 3.4 was shown to comprise of ventral extensions of the classic basal ganglia—that is, ventral striatum and ventral pallidum. Note that the part of ventral striatum shown in Fig. 3.1B is directly continuous rostrally with the accumbens-related part of ventral striatum depicted in Fig. 3.1A.

3.2.3 The Core-Shell Dichotomy

Although cytoarchitectural and histochemical features do vary to some extent between and within major parts of the striatal complex, nowhere are these differences more pronounced than in the ventral parts of the striatum, including the accumbens. Indeed, the idea of the accumbens as a well-defined functional-anatomical unit suffered another setback with the observation that it apparently comprises rather distinctly delineated core and shell subterritories (Figs. 3.1A and 3.6), which exhibit neurochemical attributes and connections that in a fundamental sense are similar (i.e., striatal in cytoarchitectural, histochemical, and connectional character) but, in the details, quite distinct (Záborszky et al., 1985; Zahm and Heimer, 1990; Brog et al., 1993). Although it is impossible to define a dorsal border between the core and the rest of the striatum, there is a more or less distinct boundary between the core and the shell. Nonetheless, it is generally understood that the shell, like the core, is an integral part of striatum. Consistent with a striatal structure, the shell is characterized by typical basal ganglia connections. However, the shell also has a number of features that are atypical for a striatal structure but that are reminiscent of the extended amygdala (see following), with which it is directly continuous. In keeping with these neuroanatomical features, the functions of the core of the accumbens are more definitively in line with those of the caudate nucleus and putamen, as in, for example, movement initiation, whereas those of the accumbens shell are less obviously involved in motor control *per*

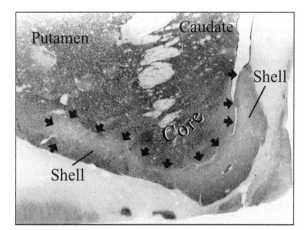

FIGURE 3.6 Photomicrograph of a calbindin immunoprocessed frontal section through the ventral striatum, including the accumbens, of the rhesus monkey. The border of the accumbens shell is marked with arrows. (Adopted from Meredith et al., 1996, reprinted with permission.)

se and more concerned with factors that underlie somatic motor function, such as motivation and the control of its vigor.

In recent years a large number of neuroscientists have based the design of their preclinical physiological and pharmacological experiments on the core-shell dichotomy. The shell, in particular, has received a lot of attention. Experimental pharmacology studies indicate that the shell is a significant target for antipsychotic drugs (Deutch and Cameron, 1992; Merchant and Dorsa, 1993), and the core-shell dichotomy plays an important role in neuronal circuit theories of addiction (see Footnote 15, Chapter 5, p. 132). The core-shell dichotomy is depicted in Fig. 3.1B, where yellow patches indicate ventral pallidal areas, and black and light blue regions denote prominent shell features, best described as small-celled islands.

3.2.4 The Heterogeneity of Ventral Striatum; Small-Celled Islands

Striatum is defined on the basis of its cortical connections, such that isocortex (neocortex) and allocortex (hippocampus and olfactory cortex) project to dorsal (caudate and putamen) and ventral striatum, respectively. However, because cortico-striatal projections from the isocortex and allocortex overlap extensively in the boundary regions between dorsal and ventral striatum, it is impossible to define a precise border between the ventral and dorsal striatum. Furthermore, projections from additional, transitional forms of cortex intermediate between allocortex and isocortex, as well as from the cortical-like laterobasal-cortical amygdala (see Chapter 4) also blend into these transitional parts of the striatum and contribute to the further degradation of boundaries between its dorsal and ventral parts.

Nonetheless, despite the gradual nature of the transitions between dorsal and ventral striatum, a large number of cytoarchitectural and histochemical features do distinguish the ventral striatum from the main dorsal parts of the caudate nucleus and the putamen. Immunoreactivity against most relevant markers has a more intense and typically blotchy pattern in the ventral striatum compared with the rest of the striatum, although the characteristic bicompartmental or so-called patch-matrix (or striosome-matrix) organization, which is revealed in the dorsal striatum with a number of histochemical methods (Graybiel and Ragsdale, 1983; Gerfen, 1992), is not readily apparent in ventral striatum. The relationship between different neurochemical markers appears to be more complex in ventral striatum than in the dorsal striatum, and this histochemical heterogeneity is matched by equally pronounced cytoarchitectonic irregularities. This is in part related to the greater tendency of neurons in ventral than dorsal striatum to form irregular aggregations. Another reason for the heterogeneity is the intermingling of large pallidal neurons with the medium-sized striatal neurons (see salmon-colored area in Fig. 3.1A).

This phenomenon becomes increasingly prominent at more caudal levels of ventral striatum (Fig. 3.1B).

Among the most intriguing contributors to the heterogeneity of ventral striatum are the small-celled islands, referred to as "terminal islands" by Sanides (1957). It is amazing to learn that the small-celled islands have generally been ignored in studies and discussions of the human brain, although they were described in great detail half a century ago. Some exceptions to this rule, however, deserve attention. Talbot et al. (1988a) published a detailed review of the islands of Calleja in different animals, and Meyer and her colleagues (1989) produced a broader study that included the entire basal forebrain in several mammals, including the human, where the small-celled islands are especially prominent. The cells of the small-celled islands are quite variable. Some of the islands contain predominantly very small "glia-like," or granular, neurons similar to those in the famous islands of Calleja of macrosmatic animals, whereas other islands, such as parvicellular islands (Fig. 3.7), contain somewhat larger neurons, many of which are still significantly smaller than the regular striatal medium-sized neurons. Small-celled islands are especially prominent in the shell but are also present in other parts of the ventral striatum and in the extended amygdala. Considering the great variability in the morphologies of individual neurons in any given cell island and the fact that the interface islands appear to be more numerous in early life, Meyer et al. (1989) suggested that transformation from one type of cell to another might take place postnatally. In spite of the fact that the small-celled islands contain a number of neurochemicals of special interest in the context of neuropsychiatric disorders (see Clinical Box 4), they are hardly ever mentioned in the clinical literature.

3.3 PARALLEL CORTICO-SUBCORTICAL REENTRANT CIRCUITS ("BASAL GANGLIA LOOPS")

As just indicated, the ventral striatum was originally defined based on its input from allocortex (olfactory cortex and hippocampus) and basolateral amygdala. However, the ventral striatum also receives prominent afferents from other nonisocortical limbic lobe regions, including the entorhinal cortex, anterior cingulate cortex, large parts of the orbitomedial prefrontal cortex, and insula (see Chapter 4). In other words, the ventral parts of the basal ganglia are related largely to allocortex (olfactory cortex and hippocampus) and other nonisocortical regions of the cortical mantle in the same way that the dorsal parts of the basal ganglia are related largely to isocortex (neocortex).

As a way of summarizing the functional-anatomical organization of the basal ganglia and its relations to the frontal cortex and the limbic lobe, it makes sense to refer to three major functional domains (Haber and Johnson

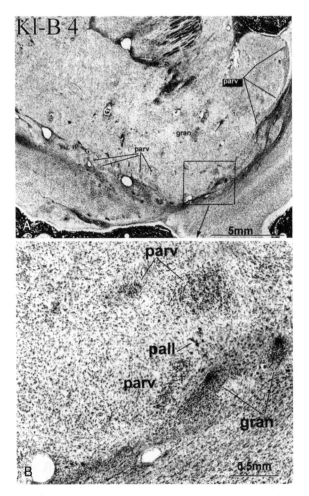

FIGURE 3.7 Photomicrographs of neurons in ventral striatum from the Klüver-Barrera stained frontal section through the human brain shown in Figure 3.2 (Kl-B4). Panel A illustrates the locations of the various types of neurons of the small-celled (interface) islands. The box in A is enlarged in B. Although Klüver-Barrera sections are not optimally suited for cytoarchitectural studies, they do provide a clear picture of the various cell types that intermingle with striatal neurons in the accumbens. Cells belonging to granular (gran) and parvicellular (parv) interface islands are shown. A group of larger, likely pallidal, neurons (pall) is also shown in B. (Modified from Heimer et al., 1999, reprinted with permission.)

CLINICAL BOX 4

Small-Celled Islands: Clinical-Anatomical Correlations

With the increasing awareness of the important consequences of neuroplasticity as a prominent variable, not only in normal development but also in the pathophysiology of major psychiatric disorders (Schneider, 1979; Stevens, 1992, 2002; Mesulam, 2000; Duman et al., 2000), the small-celled islands, which are especially conspicuous in the ventral striatum, may deserve considerably more attention than they have enjoyed in the past.

The neurons in the small-celled islands contain an abundance of opioid and dopamine receptors (Voorn et al., 1996) in addition to Bcl-2 protein (Bernier and Parent, 1998), a marker of neuronal immaturity. When Sanides (1957) suggested that the small-celled islands might contain progenitor cells arrested in development (hence the term *terminal islands*), the notion of plasticity (although proposed by Ramòn y Cajal more than a century ago), not to mention postnatal neuronal development (Eriksson et al., 1998), was not part of mainstream thinking. However, when one considers the potential for development of these neurons, together with the fact that they appear to be more numerous in early life (Sanides, 1957; Meyer et al., 1989), it is not difficult to envision the ventral striatum as especially prone to remodeling in response to changing circumstances.

Cortical neurogenesis in adult primates including the human (Eriksson et al., 1998; Gould et al., 2006) is now generally accepted, and a recent study of neurogenesis in the basal ganglia of primates (Poloskey and Haber, 2005) is highly relevant in this context. As indicated by Poloskey and Haber, it may be important in this context that the ventral striatum, and especially its shell component, is in the direct line of the so-called "rostral migratory stream" originating in the subventricular zone of the lateral ventricle. Acknowledging the extended column of parvicellular neurons alongside the medial edge of the ventral striatum (Fig. 3.7), it is tempting to imagine how developing neurons might stream down from the lateral ventricle to the more ventral parts of the striatal complex. Be that as it may, the point is powerfully made by the above-mentioned study, that among the various striatal regions, the most dramatic variation in the number of proliferating cells from birth to adolescence, takes place in the ventral striatal regions related to orbital and ventromedial prefrontal cortical regions, known to be of importance for emotional–motivational functions. If we submit to the idea of neuropsychiatric disorders as "neuronal circuit dysfunction disorders," it is tempting to suggest some kind of relationship, as did Poloskey and Haber, between plasticity in ventral striatum and development of neuropsychiatric disorders. Such disorders typically emerge during development, particularly during adolescence or early adult age—in other words, at a time when the corticosubcortical reentrant circuits through the ventral parts of the basal ganglia appear to be especially vulnerable to extrinsic or intrinsic factors during postnatal development (Heimer, 2000; Stevens, 2002). The ventral parts of the basal ganglia are often mentioned as critical regions for transmitter interactions in schizophrenia (Gurevich, 1997; West and Grace, 2001). Apropos the dopamine hypothesis of schizophrenia, which was mentioned earlier in this chapter, the apparent elevation of dopamine D-3 receptors in ventral striatum (Fig. A) in drug-free schizophrenia patients (Gurevich et al., 1997) presents an intriguing clinical-anatomical correlation, especially since the D-3 receptor may be a potential site for genetic polymorphisms related to increased susceptibility for schizophrenia. These and other interesting questions, including, for example, the relationship between drug abuse and schizophrenia, and the potential treatment of emerging schizophrenia with D-3 receptor antagonists are discussed in a recent paper by Joyce and Millan (2005). As mentioned earlier, the shell of the ventral striatum, where the majority of the small-celled islands are located, appears to be an especially significant target for antipsychotic drugs, and several studies also point to the shell as an especially important region for interactions between dopamine D-3 receptors and drugs of abuse. The conundrum posed by the coexistence of organizational and neurochemical similarities together with functional differences when considering the shell of the ventral striatum and extended amygdala and the potential involvement of these macrosystems in drug abuse is discussed in Footnote 15, Chapter 15, p. 132.

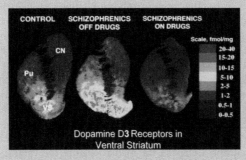

CBOX 4 FIGURE A   PET imaging demonstrating elevated density of dopamine D3 receptors in the ventral striatum of schizophrenics when off antipsychotic drugs as compared to controls is reduced by antipsychotic drug treatment. (Reprinted from Gurevich et al., 1997, with permission.) See color plate.

Gdowsky, 2004). This idea, which reflects the close cooperation between the basal ganglia and the frontal lobe in the execution of goal-directed behaviors, is schematically illustrated in Fig. 3.8, in which the colors denote functional distinctions—that is, blue for motor functions, yellow for cognitive-executive functions, and red for emotional-motivational functions. These functional domains are broadly related to motor cortical areas, prefrontal isocortical areas, and limbic lobe regions, respectively. The separateness of functional domains, however, is blurred because the corticostriatal projections from adjacent functional domains within the striatum overlap extensively with each other, as indicated by the gradual changes of color in Fig. 3.8. In other words, just as the ventral striatopallidal system cannot be strictly separated from

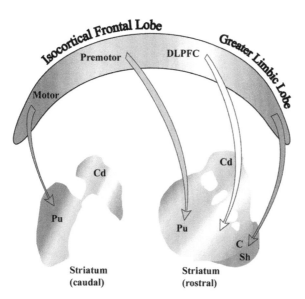

FIGURE 3.8 Diagram of corticofugal projections from the isocortical frontal lobe and the greater limbic lobe of the human brain. The gradual changes in color indicate overlapping projections. Although the crescent-shaped figure represents the isocortical frontal lobe and the nonisocortical greater limbic lobe only, it should be emphasized that cortical-striatal projections do come from the entire cerebral cortex. The greater limbic lobe regions include all nonisocortical parts of the cerebral cortex, including the cortical-like laterobasal-cortical amygdala (see Figs. 4.1 and 4.4). (Original graphic artwork by Suzanne Haber, modified from Heimer, 2003, with permission.) See color plate.

the rest of the basal ganglia, the striatal functional domains cannot be cleanly separated from each other. Limbic lobe projections (related to emotional-motivational functions) overlap with dorsolateral prefrontal projections (serving cognitive-executive functions) in the central part of the striatal complex (see shades of orange in Fig. 3.8). Dorsolateral prefrontal projections to striatum, likewise, overlap with projections from motor cortical areas (see shades of green). Moreover, these projections also are governed by a principle described by Yeterian and Van Hoesen in 1978, who showed that functionally related, reciprocally interconnected parts of the cortex project in part to the same foci within the striatum.

The discovery of the ventral striatopallidal system and revelation that the ventral striatopallidal system projects to the mediodorsal (MD) thalamus (Heimer, 1978; Heimer et al., 1982; Young et al., 1984) rather than to the globus pallidus-related ventral anterior–ventral lateral (VA–VL) "motor" nucleus of the thalamus, marked the beginning of the notion of "parallel" cortico-subcortical reentrant circuits, also known as "frontal-subcortical circuits" or "basal ganglia loops" (Fig. 3.9), which has been reviewed and amplified in numerous papers (Alexander et al., 1986, 1990; Groenewegen et al., 1990, 1994; Joel and Weiner, 1994; Zahm and Brog, 1992). A cortico-basal

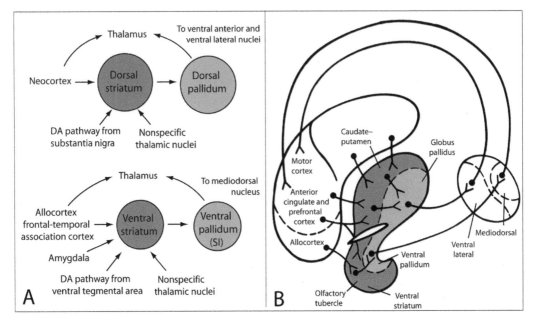

FIGURE 3.9 Dorsal and vental cortico-subcortical circuits through the basal ganglia and thalamus. The images show basic similarities between the classic dorsal and ventral cortico-striatopallido-thalamic reentrant circuits. A. After the discovery of the ventral striatopallidal system (Heimer and Wilson, 1975), Heimer (1978) proposed that the mediodorsal nucleus, rather than the ventral lateral thalamic motor nucleus, is the likely thalamic target for the ventral striatopallidal system. B. This image emphasizes the parallel character of the cortico-subcortical reentrant circuits throughout the dorsal and ventral regions of the basal ganglia in the rat. (Reprinted from Heimer, 2003, with permission.)

ganglia-thalamocortical reentrant circuit comprises, in sequence, corticostriatal, striatopallidal, pallidothalamic, and thalamocortical links, with a defining characteristic being that the final thalamocortical link returns to the same area of the cortex from where the initial corticostriatal link originated—that is, the sequential links form a true circuit. (And, by the way, where the circuit actually begins and ends cannot be unequivocally stated, although conventionally the beginning is typically thought to be in the cortex. The very interesting work of Earl Miller and colleagues [Pasupathy and Miller, 2005; see also following and section 5.3.4], however, suggests that this may not necessarily, or at least not always, be the case.)

Six major frontal-subcortical circuit categories are usually recognized (Alexander et al., 1990; Middleton and Strick, 2001), two of which are related to motor cortical areas (a motor circuit originating in the supplementary motor cortex and an oculomotor circuit originating in the frontal eye field), and four of which relate to prefrontal cortical regions (an executive circuit originating in the dorsolateral prefrontal cortex, an anterior cingulate circuit originating in the anterior cingulate cortex, and medial and lateral orbitofrontal circuits with origin in the medial and lateral orbital cortex, respectively). Three of these circuits—the anterior cingulate circuit, the lateral orbitofrontal circuit, and the medial orbitofrontal circuit—are considered part of the ventromedial emotional-motivational domain of the striatum (red-orange in Fig. 3.8) and more or less correspond to the ventral striatopallidal system originally described in the rat. A consensus seems to have emerged that the cortico-subcortical reentrant circuit (and its subcircuits) through the ventral striatopallidal system, popularly referred to as the "limbic loop" (the "motive" circuit of Kalivas et al., 1999) is critical for selecting and mobilizing appropriate adaptive behavior ("reward-guided choice behavior" of Schultz et al., 2000), in the same fashion that the executive circuit (and its subcircuits) is implicated in action planning and working memory, or the "motor" circuit in the execution of motor functions. While true that the notion of frontal-subcortical circuits has been eagerly embraced by clinical neurologists and psychiatrists (Clinical Box 5), the science of these circuits is still in its infancy. Circuits in addition to the frontal-subcortical ones just mentioned appear to involve both the temporal and the parietal lobes (Middleton and Strick, 2001), and more subcircuits will no doubt be identified within the broader category of circuits just mentioned (see Chapter 4). Although frontal-subcortical circuits are currently promoted as principal organizational networks in the brain, many important questions await answers. For instance, how many circuits can reasonably be defined within each of the major functional domains, and to what extent, and how, do the various circuits interact?

To summarize, the ventral striatopallidal system is related primarily to nonisocortical parts of the cerebral cortex, including significant parts of the orbitomedial, cingulate, and insular cortices, as well as medial temporal lobe structures like hippocampus, laterobasal amygdala, and surrounding cortical areas. The ventral striatopallidal system and its thalamic projection to the MD-prefrontal system served as a catalyst for the development of a more global appreciation of cortico-basal ganglia-thalamocortical reentrant circuits. The structures involved in these circuits, usually recognized as "limbic" in the neuroscience literature, are often the focus of pathologic changes in several of the most devastating neuropsychiatric disorders. As a consequence, the "basal ganglia loops" have become the subjects of numerous research programs with the goal of identifying the functional-anatomical substrates for various neuropsychiatric disorders (Clinical Box 5). This trend is reflected in the publication of a large number of clinically oriented basic science papers during the last 10–15 years (e.g., Swerdlow and Koob, 1987; Robbins, 1990; Cummings, 1993; Deutch, 1993; Joyce, 1993; Mega and Cummings, 1994; Groenewegen, 1996a; O'Donnel and Grace, 1998; Baxter, 1999; Joyce and Gurevich, 1999; Heimer, 2000).

CLINICAL BOX 5

Reentrant Circuits in a Clinical Context

The concepts of the ventral striatopallidal system and parallel cortico-subcortical reentrant circuits have been tied to efforts to explain different neuropsychiatric disorders (Lichter and Cummings, 2001). Thus, various symptoms are often related to neuronal circuits through the ventral parts of the basal ganglia, due to the perceived importance of these circuits for emotional and motivational functions. As should have been clear from the description of the origin and development of our knowledge of these circuits, they are central not only to motor systems but also to the behavioral ramifications of emotional function and dysfunction. For some time, neurologists have concluded that dysregulation of the cascade of inhibitory and excitatory neurotransmitters that regulate activity in the cortico-basal ganglia-thalamocortical loops disrupts motor output, as in, for example, the tremor in Parkinson's disease. However, we also have come to view emotional expression as similarly dependent on the function of analogous motor output systems in the basal forebrain—that is, involving the ventral striatopallidum and extended amygdala. These considerations are underpinned by the "new neuroanatomy" of the limbic forebrain.

Although these ideas, which have arisen out of basic neuroanatomical and neurophysiological experiments, are still too simplistic to explain many clinical observations, they have guided clinical thinking as regards neuropsychiatric symptomatology and treatment strategies, not the least of which include neurosurgical interventions for neurological and behavioral disorders. In general, the parallel distributed nature of the cortico-subcortical reentrant circuits suggests that a lesion in one circuit will be somewhat self-contained, and that the resulting behavioral effects will be different to those noted with changes in an adjacent circuit. A second implication is that intervention or lesion at any of a number of nodal points within a given circuit (e.g., the pallidum) will lead to similar effects on the clinical picture. This is illustrated for example by the description of three different prefrontal syndromes, noted following lesions within the three described prefrontal circuits.

Central to the three syndromes attributed to prefrontal cortical-subcortical circuit dysfunction (involving the ventral striatum) is alteration of emotional and motivational behaviors. In contrast, in Parkinson's disease there is loss of dopaminergic input to the putamen and dorsolateral caudate-putamen—that is, the dorsal striatal circuitry. Accordingly, this disorder initially presents with motor symptoms, some tremor, and muscle stiffening. However, as the disease progresses or is influenced by treatment, it comes to involve ventral striatal structures, whereupon behavioral problems then also emerge as a significant part of the clinical picture. So, in Parkinson's disease, 40 percent of patients will develop depressive syndromes, often a dysthymia with a predominant anxiety, or a major depressive disorder. As the disease progresses, psychoses may appear. It is estimated that 30 percent of patients present with hallucinations, and about 10 to 20 percent develop frank psychoses often exacerbated by dopamine agonists.

The hallucinations are predominately visual, complex, and are often related to complete scenes of events, such as, for example, with the patients seemingly interacting with many former colleagues or friends outside the hospital, who are greeting or persecuting them. The associated delusions are usually paranoid and commonly nocturnal. Patients may show classic sundowning—in other words, as the sun goes down, the patient's psychosis lights up.

A variant of the behavioral syndromes associated with these motor-motivational circuits is the dopamine dysregulation syndrome, where a patient deliberately overdoses on antiparkinsonian medication to obtain a euphoric effect but often with disastrous consequences. The features of this syndrome are shown in Table 1.

In psychiatric context, the cortico-subcortical circuits have been used to help us gain insight into the underlying pathophysiology of several disorders. A relevant paradigm in this regard is obsessive compulsive disorder (OCD). This neurosis in many cases seems very neurological and in severe forms can present with mainly motor signs, such as in extreme motor slowness. Many cases of secondary OCD are reported in the literature with the main involved brain structures being the frontal cortex, basal ganglia, and thalamus (Trimble, 1996). There is frequent comorbidity of OCD with several basal ganglia disorders such as Sydenham's chorea, Gilles de la Tourette's syndrome, frontotemporal dementia, and some encephalitides, such as encephalitis lethargica. Further, follow-up studies indicate that children presenting with OCD almost invariably go on to develop tics. The behavioral observations have been accompanied by detection of alterations in frontal-basal ganglia circuitry using brain imaging techniques (Rauch et al., 1997). The idea that has emerged is that OCD reflects dysfunction in the reentrant cortical-subcortical circuits, with decreased pallidal inhibition resulting in thalamocortical excitation. This suggestion is supported by a long history of the use of lesions in the basal ganglia circuitry in the management of such conditions. In fact, neurosurgery presently is used for OCD only in extreme cases, although the more recently developed brain stimulation techniques promise to become a more popular approach.

CBOX 5 TABLE 1

Dopaminergic Dysregulation

- overuse of dopaminergic medication
- risk factors: sex (male); age (young)
- increasing doses of dopaminergic drugs despite severe dyskinesias
- behavioral and mood disturbances
  - hypomania-mania
  - irritability, aggression
  - paranoia
  - punding (repetitive, purposeless motor acts)
  - walking aimlessly
  - pathological gambling
  - drug hoarding
  - hypersexuality

3.4 THE EXTENDED AMYGDALA

The extended amygdala is another basal forebrain functional-anatomical macrosystem that is slowly making its way into the general neuroscience textbooks.
The concept of the extended amygdala originated with the rediscovery of
pioneering comparative and developmental studies by J. B. Johnston (1923),
who noted in studies of human embryos and lower vertebrates that the centromedial amygdala and bed nucleus of the stria terminalis form a continuum.
Johnston also realized that columns or islands of cells forming a partly interrupted continuum are still evident in the stria terminalis in adult mammals,
including humans. But Johnston's groundbreaking work was soon forgotten,
until it was rediscovered and amplified by de Olmos (de Olmos and Ingram,
1972), who identified a histochemically distinct continuum in adult mammals
comprising the centromedial amygdala, bed nucleus of stria terminalis, and
similarly organized neural tissue extending between them within the sublenticular (subpallidal) part of the basalis region. The rediscovery of the extended
amygdala is reviewed in Basic Science Box 3, and its importance as an output
channel for the greater limbic lobe will be discussed in Chapter 5. Suffice it
to say, as in the case of the ventral striatopallidal system, that the groundbreaking discovery of the subpallidal neuronal continuum between the amygdala
and the bed nucleus of stria terminalis was made possible by a newly developed
silver method—in this case, that of Jose de Olmos (1969), which for some
reason has a specific chemical affinity for part of the extended amygdala.

The extended amygdala has many distinct anatomical and histochemical
characteristics, which distinguish it from surrounding areas and which have
served as a basis for its identification in the brain of a large number of
mammals, including the human (de Olmos, 1985, 2004; Martin et al., 1991;
Heimer et al., 1999). Whereas the ventral striatopallidal system occupies the
main rostral and subcommissural parts of the basalis region (Figs. 3.1A and
B), the components of the extended amygdala, which are colored yellow and
green in Fig. 3.1, are located more caudally, in the sublenticular part of the
basalis region (Figs. 3.1C and D).

The best way to appreciate the three-dimensional structure of the extended
amygdala is to imagine it isolated from the rest of the brain, as a ring formation encircling the internal capsule (Figs. 3.1D and 3.10). The term *extended
amygdala* emphasizes the fact that the central and medial nuclei of the amygdala in the temporal lobe (Fig. 3.1D, Ce and Me in Fig. 3.10) are continuous
rostrally and medially with the lateral and medial parts of the bed nucleus of
the stria terminalis (Figs 3.1B and C, BSTL and BSTM in Fig. 3.10). The centromedial amygdala and the bed nucleus of the stria terminalis are joined by
cell groups or columns of neurons both alongside the stria terminalis (supracapsular bed nucleus of stria terminalis in Fig. 3.1D, BSTS/st in Fig. 3.10)
and beneath the internal capsule and globus pallidus (sublenticular extended

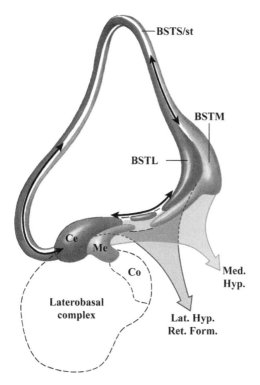

FIGURE 3.10 The extended amygdala (yellow and green) shown in isolation from the rest of the brain, with the extensions of the central (Ce) and medial (Me) amygdaloid nuclei within the stria terminalis (st) and through the sublenticular region to the bed nucleus of stria terminalis (BST). The central division of the extended amygdala is coded in yellow and the medial division in green and the major outputs from each are indicated by the respective yellow and green arrows. The black arrows represent the abundant long and short intrinsic associational axonal connections that characterize the macrosystems. Note that the laterobasal-cortical amygdaloid complex, which is included in the concept of the limbic lobe, is not part of the extended amygdala. Further abbreviations: BSTL—lateral bed nucleus of the stria terminalis; BSTM—medial bed nucleus of the stria terminalis; BSTS/st—supracapsular part or the bed nucleus of stria terminalis; Co—cortical amygdaloid nucleus. (Modified from Heimer et al., 1999, with permission; art by Medical and Scientific Illustration, Crozet, VA.) See color plate.

amygdala, Fig. 3.1C, SLEA in Fig. 3.10). The discovery of the extended amyg-dala is reviewed in Basic Science Box 3.

As indicated in the introduction to this chapter, one of the most obvious consequences of this paradigm shift in thinking about amygdalar organization is that the amygdala cannot be conceived of as a well-defined, uniform func-tional-anatomical unit restricted to the temporal lobe. To the contrary, it is absolutely essential to understanding the concept of the extended amygdala to realize that the cortical-like laterobasal complex and cortical nucleus of the

BASIC SCIENCE BOX 3

Discovery of the Extended Amygdala

As just indicated the idea of a centromedial amygdala-bed nucleus of stria terminalis continuum, which originated with J. B. Johnston's pioneering studies almost a century ago, was quickly forgotten, in all likelihood because it proved difficult to identify and illustrate the attenuated cellular continuities between the centromedial amygdala and the bed nucleus of stria terminalis. An additional breakthrough was needed, and, as is often the case in neuroanatomy, it came with a histotechnical advance—in this case, in the form of de Olmos's (1969) cupric silver method. For some unknown reason the cupric-silver method exhibits chemical specificity for a special type of neuron and neuropil, restricted to the central amygdaloid nucleus, the lateral bed nucleus of stria terminalis, and identical neuronal elements bridging the gap between these two structures, not only in the sublenticular region as shown in Fig. A but also in the stria terminalis.

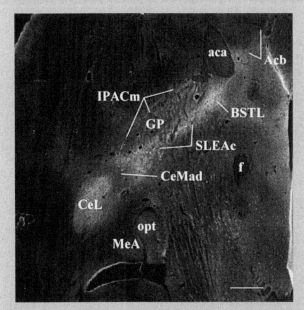

BSBOX 3 FIGURE A   Dark-field image of a section from a rat brain cut in the horizontal plane and processed with normal cupric silver staining showing the central division of the extended amygdala. When this method is applied to unlesioned brains, these argyrophilic normal fibers are the last to be suppressed by preincubation or removed by subsequent bleaching. Abbreviations: Acb—accumbens; aca—anterior limb of the anterior commissure; BSTL—lateral division of the bed nucleus of the stria terminalis; CeMad—medial division of the central amygdaloid nucleus, anterodorsal; CeL—lateral division of the central amygdaloid nucleus; f—fornix; GP—globus pallidus; IPACm—medial division of the interstitial nucleus of the posterior limb of the anterior commissure; MeA—medial amygdaloid nucleus; opt—optic tract; SLEAc—sublenticular extended amygdala, central division. (Reprinted from Alheid et al., 1995, with permission.)

The sublenticular and supracapsular (within the stria terminalis) parts of the extended amygdala were subsequently detected in a number of mammals including the monkey and the human, using various tracer techniques and immunohistochemical methods (Alheid and Heimer, 1988; Alheid et al., 1995; Heimer et al., 1999; Martin et al., 1999; Alheid, 2003; McDonald, 2003; de Olmos, 2004). An impressive panoramic view of the sublenticular extended amygdala, illustrated in Fig. B, has been obtained in the monkey using autoradiography following injections of a radioactive tracer in the centromedial amygdala (see also Amaral et al., 1989).

Although the existence of a sublenticular extended amygdaloid continuum in the human has been described by several authors (de Olmos et al., 1985), it has been more difficult to obtain a panoramic view of this comparable to that

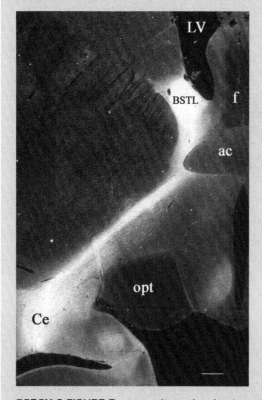

BSBOX 3 FIGURE B Autoradiography of amino acid transport after an injection in the central nucleus (Ce) of the macaque amygdala (dark-field illumination). Note the dense labeling of axons and terminals in the sublenticular extended amygdala. This high degree of associative connections distinguishes the extended amygdala from the overlying striatopallidal complex. Abbrevations: ac—anterior commissure; BST—bed nucleus of the stria terminalis; GP—globus pallidus; opt—optic tract. (Reprinted from Alheid et al., 1990, with permission.) See color plate.

obtained in the monkey, where the central extended amygdaloid continuum is nicely exposed in regular coronal sections. This is in part because the sublenticular extended amygdaloid components in the human are more widely separated by fiber bundles, as is reflected in the schematic drawing of the isolated extended amygdala in Fig. 3.10.

A somewhat similar situation exists in regard to the extended amygdaloid continuum within the stria terminalis—that is, the supracapsular extended amygdala. Many scientists, starting more than a century ago with Ziehen (1897), have described single cells and small clusters of cells in different parts of the stria terminalis in many mammals, including the human (Sanides, 1958; Strenge et al., 1977; Heimer et al., 1999), but it has been difficult to illustrate the extended amygdaloid continuum in the stria terminalis in a panoramic fashion. However, Alheid et al. (1998) were finally able to do this using immunohistochemical detection of MAP-2 (microtubule-associated protein type 2) in sections cut parallel to the plane of the stria terminalis. In these preparations, continuous columns of exended amygdaloid neurons, represented largely by the MAP-2 immunoreactive dendrites, were detected in the stria terminalis of the rat (Fig. C).

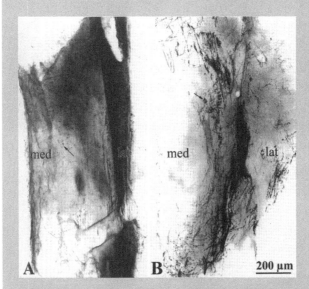

BSBOX 3 FIGURE C Microtubule-associated protein 2 immunostained horizontal (A) and frontal (B) sections through the dorsally convex and retrocapsular parts, respectively, of the stria terminalis. Note the concentration of longitudinally oriented dendritic processes in the lateral part of the stria. However, several dendrites can also be seen to cross more or less transversely through the stria. Images of neuronal cell bodies are clearly displayed, especially among the dendritic bundles in the lateral (lat) and medial (med) pockets, but also in other parts of the stria (arrows). (Reprinted from Alheid et al., 1998, with permission.)

There is no denying the relative attenuation of the preceding continuums by comparison to the two major parts of the extended amygdala: the bed nucleus of stria terminalis and the centromedial amygdala. Nonetheless, these neuronal columns, however attenuated, evoke the image of a continuous ring-like structure, as illustrated in Fig. 3.10. Furthermore, it is important to realize that the existence of the extended amygdala as a macroanatomical basal forebrain structure, distinct from the striatopallidal system, is also based on additional criteria, including developmental, connectional, and histochemical (see previous references cited). Notwithstanding some transitional areas between the striatopallidal system and the extended amygdala—the caudomedial shell region of the accumbens being a prime example—prominent developmental, connectional, and histochemical features distinguish these two macrosystems, and the differences are bound to have functional consequences.

amygdala (Fig. 3.10) are not part of the extended amygdala. This important consideration applies not only to the notion of the extended amygdala but also to fully appreciating the concept of the "greater limbic lobe," (to be discussed in Chapter 4), within which are included the laterobasal complex and cortical amygdaloid nucleus but not the centromedial amygdala. These fundamental organizational principles are consistent with Brockhaus's original subdivision of the amygdala into an "amydaleum proprium" and "supra-amygdaleum" (Brockhaus, 1938), and, likewise, they are consistent with the prescient work of Johnston (1923). Clearly, the concept of the extended amygdala was foreshadowed in the descriptions of the human brain by these pioneering neuroanatomists (Heimer et al., 1999).

The amygdala is widely recognized for its importance in emotional processing (Klüver and Bucy, 1937; Weiskrantz, 1956)—that is, for evaluating sensory stimuli in terms of their emotional and motivational significance (Gloor, 1997). The laterobasal-cortical amygdala, in particular, is characterized by multisensory input from higher-order cortical association areas and is therefore in an excellent position to serve as a "sensory-emotional" interface (Zald, 2003). The extended amygdala, in turn, receives much of its input from the laterobasal amygdala but also from other parts of the limbic lobe, and it is characterized by projections (Fig. 3.10) to hypothalamic and brainstem autonomic and somato-motor structures (primarily from the central extended amygdala in yellow color) and to the endocrine-related medial hypothalamus (primarily from the medial extended amygdala in green color). Thus, the extended amygdala, apparently, is a complex of effector systems—that is, "emotional-motor" interfaces, for some of the emotional circuits in the brain (Alheid and Heimer, 1988; Heimer et al., 1997). The prevailing view in the past had been that the bed nucleus of stria terminalis is above all a relay in the downstream projections from the centromedial amygdala. In the concept of the extended amygdala, however, the bed nucleus of stria terminalis and the centromedial amygdala are part of one and the same macrostructure and the outputs from both, and the intervening parts of the sublenticular region, are functionally comparable. Since extended amygdala is also characterized by very dense, intrinsic (short and long associative) connections, it represents a strategically placed effector complex capable of coordinating activities in multiple limbic lobe regions for the development of behavioral responses through its output channels (Alheid and Heimer, 1988; Heimer et al., 1999).

Extended amygdala is a hotbed for neuropeptides and neuropeptide receptors. Receptors (e.g., for vasopressin, oxytocin, androgens) and neuropeptides (e.g., cholecystokinin, angiotensin, somatostatin, neurotensin, opioid peptides) that have attracted special attention in neuropsychiatric disorders are highly characteristic for one or both of its divisions. It is also worth mentioning that CRF (corticotrophin-releasing factor) containing neurons, which are especially prominent in the central division of the extended amygdala, are highly relevant in physiologic responses to stress, which is generally considered to be an important contributing factor in many neuropsychiatric disorders, including schizophrenia, major depression, and anxiety disorders (see Basic Science Box 4).

Thus, a large number of intriguing functional-anatomical and pharmacological correlations support the potential relevance of the extended amygdala in a variety of neuropsychiatric disorders. Scientists advancing theories of drug abuse and addiction have focused their interest on the large forebrain continuum represented by the extended amygdala and the shell of the accumbens (Koob, 1999). The recent focus on the bed nucleus of stria terminalis and its potential role in fear, stress, and anxiety (Walker et al., 2003), furthermore, is likely to facilitate the acceptance of the extended amygdala as a theory of great practical and heuristic value in the appreciation of emotional behavior and neuropsychiatric disorders.

3.5 SEPTAL-PREOPTIC SYSTEM

Whereas the striatopallidum and extended amygdala have generated considerable interest among neuroscientists and have been amply validated as functional-anatomical entities (e.g., Newman and Winans, 1980; Haber et al., 1985; Koob, 1999; McDonald, 2003; Morelli et al., 1999; Pinna and Morelli, 1999), including in the primate (e.g., Martin et al., 1991; Meredith et al., 1996; Voorn et al., 1996; Haber and Gnodski, 2004), a macrosystem posed as centered on the lateral septum and diagonal band by Alheid and Heimer (1988) and Swanson (2000) is otherwise less well represented in the literature. Characterization of the septum as representing a distinct basal forebrain macrosystem has not been disputed, however, and the lack of a substantial

BASIC SCIENCE BOX 4

The Extended Amygdala: A Contested Concept

The introduction of the concept of the extended amygdala has not proceeded entirely without resistance. Indeed, whether it is a useful idea at all was propelled into a heated debate upon the publication of a paper by Larry Swanson and his colleagues (Canteras et al., 1995). Although this debate was renewed many times subsequently (Heimer, 1997; Swanson and Petrovich, 1998; de Olmos and Heimer, 1999; Dong et al., 2001; Swanson, 2003; Alheid, 2003; Zahm et al., 2003), it seems appropriate to conclude this review of the extended amygdala with a summary of the major arguments. However, one might have thought that much of the steam would have dissipated from the controversy, since, as McDonald (2003) has indicated, "Golgi, immunohistochemical and tract tracing investigations conducted during the 1980s and the late 1970s corroborated Johnston's concept that the central and medial nuclei (of the amygdala) do extend rostromedially to become continuous with the BST (bed nucleus of stria terminalis)." In other words, if it is deemed useful to keep the terms *central* and *medial amygdaloid nuclei* as part of the neuroanatomical nomenclature (and everybody, including Swanson and his colleagues, do seem to use these terms to their advantage), and if everybody agrees with Johnston and McDonald that the central and medial amygdaloid nuclei extend rostromedially to become continuous with the bed nucleus of stria terminalis, the controversy surrounding the choice of the term *extended amygdala* for this continuum appears at first glance to be exaggerated.

According to Swanson and his collaborators, the centromedial amygdala and bed nucleus of stria terminalis are integral parts of the striatopallidal system, with centromedial amygdala representing a striatal structure and the bed nucleus of stria terminalis its pallidal target. As indicated by Alheid (2003), it may be appropriate to ask "what's in the name." Does it make sense, as suggested by Swanson (2003), to regard the centromedial amygdala as a striatal structure and the bed nucleus of stria terminalis as pallidum but qualify the viewpoint with the caveat that "limited regions of the former may be pallidal, and limited regions of the latter may be striatal"? Why not, then, just accept the centromedial amygdala-bed nucleus of stria terminalis continuum as a macroanatomical system in its own right? Both Alheid and Swanson reminded us of the fact that extensive morphological and histochemical similarities in regard to neuronal units and some of the major connections characterize all of the topographically distinct cortico-subcortical telencephalic systems. Alheid had already eloquently drawn attention to these similarities in 1988 (Alheid and Heimer, 1988) when he also included the septum-diagonal band nuclei in this comparison. Zahm (2003, 2006) has recently referred to these basic similarities as the "generalized corticofugal template," based on the original description of Alheid and Heimer (1988).

There is also general agreement that there are major differences between neurochemical and neuroanatomical organizations of the ventral striatopallidal system and the extended amygdala. Whereas the striatopallidum, including the ventral striatopallidal system, is widely known for its involvement in parallel cortico-subcortical reentrant circuits (Heimer et al., 1982; Alexander et al., 1986; Groenewegen et al., 1990; Zahm and Brog, 1992; Joel and Weiner, 1994), one of the distinguishing features of the extended amygdala is the massiveness of its system of internal associative connections (Alheid and Heimer, 1988). This and other characteristic features of the extended amygdala, which have been prominently reported in many publications, including a recent series of papers by Shammah-Lagnado et al. (Shammah-Lagnado, 1999; Shammah-Lagnado et al., 2000, 2001), provided the anatomical basis for the suggestion that the extended amygdala is well suited to coordinate activities of the many forebrain regions that project to this strategically placed ring formation (Heimer et al., 1999).

As regards long intrinsic connections, Swanson (2000) refers to a paper by Usuda et al. (1998), which, like a few other papers (Haber et al., 1990; Heimer et al., 1991; Van Dongen et al., 2005), has described intrastriatal association fibers. However, the relative scarcity and limited reach of such intrastriatal connections, compared to the dense and pervasive associative system within the extended amygdala, seem more like "exceptions that prove the rule"—in other words, that the striatopallido-thalamic projection systems are above all characterized by a "narrow topography" (Alheid, 2003).

As for histochemical differences between the ventral striatopallidal system and the extended amygdala, a large number of receptors and neuroactive substances that characterize one or both subdivisions of the extended amygdala (central and medial) are more or less absent in the ventral striatopallidal system. Especially significant in this context is the nearly complete absence of parvalbumin immunoreactive interneurons in the extended amygdala, in comparison to their common occurrence in the striatum, including the ventral striatum (Zahm et al., 2003). Such neurons are a critical determinant of striatal electrophysiology (Koos and Tepper, 1999).

Swanson and his colleagues (Alvarez-Bolado et al., 1995; Swanson, 2000, 2003) have argued that the centromedial amygdala and bed nucleus of stria terminali are derived from the lateral ganglionic eminence (their "striatal" ridge) and medial ganglionic eminence (their "pallidal" ridge), respectively. Our published commentary on this issue (Heimer et al., 1997), however, noted that Song and Harlan (1993, 1994) and Bulfone et al. (1993) report a common origin for much of the extended amygdala, a notion that received additional support from another developmental study (Nery et al., 2002) in which the authors conclude that the centromedial amygdala, bed nucleus of stria terminalis and caudal accumbens derive from a caudal ganglionic eminence, that they regard as distinct from both the medial and the lateral ganglionic eminence. This study, likewise, revealed that only a very small percentage of the cells derived from what they refer to as the caudal ganglionic eminence (the apparent origin for the extended amygdala) express parvalbumin, which is consistent with Zahm's observation discussed in the previous paragraph.

Considering the above-mentioned differences between the extended amygdala and the ventral striatopallidal system, and with special consideration of the above-mentioned developmental studies, it is difficult to argue that these differences are insignificant. But the fact remains—and this point has been acknowledged on both sides of this controversy (Alheid in 1988, Swanson in 2000)—that the global structure of the cortico-subcortical telencephalic relationships (the "generalized corticofugal template") is similar for striatopallidum, extended amygdala, and septal-preoptic system (see Chapter 5). It is also safe to conclude that this dispute will serve to facilitate our understanding of this part of the brain in the sense that the debate will likely stimulate the execution of additional studies designed to reveal additional features of these macrosystems, as well as further differentiations and subdivisions within the systems. In the meantime, it appears that a growing number of scientists have accepted the extended amygdala, for example, as a useful concept in investigations of drug addiction (Koob and Le Moal, 2006) and several other specific emotional and motivational behaviors ranging from fear and anxiety to sexual and appetitive behavior (McGinty, 1999).

supporting literature is in all likelihood related more to the fact that few studies have focused on the functional-anatomical organization of the septum, rather than any real opposition to the framework.

Medium spiny neurons (Alonso and Frotscher, 1986) that get substantial inputs from hippocampus, ventral mesencephalon, and intralaminar thalamus (reviewed in Jakab and Leranth, 1995, and Swanson et al., 1987) are prominent in the lateral septum. Moreover, in addition to its well-known projections to the vicinity of the magnocellular corticopetal cholinergic neurons in the medial septum-diagonal band (Leranth and Frotscher, 1989), the lateral septum also projects strongly to the preoptic area and rostral lateral hypothalamus (Jakab and Leranth, 1995), which contain abundant medium-size, sparsely spined GABAergic neurons that project to the brainstem and thalamus (Swanson, 1976; Swanson et al., 1984, 1987). These features conform in all major respects to the generalized template for basal forebrain macrosystems. The lateral septum and preoptic region, considered together as a forebrain macrosystem, will be referred to in this book as the "septal-preoptic system."

3.6 THE MAGNOCELLULAR BASAL FOREBRAIN SYSTEM (BASAL NUCLEUS OF MEYNERT)

Cellular components of the extended amygdala interdigitate with major neuronal aggregations representing the basal nucleus of Meynert, which is part of the so-called magnocellular basal forebrain system, so named because many of its cells are larger than cells in the surrounding areas. Since many of these strikingly large hyperchromatic cells are located within the area that used to be referred to as the substantia innominata, there has been a tendency to use the basal nucleus of Meynert as a synonym for substantia innominata. It had been difficult to take issue with this practice, which was common throughout the last century, since little was known then about the other cell populations in the basalis region. But that situation has now changed, insofar as we recognize that much of substantia innominata is occupied by ventralward incursions from ventral striatum and extended amygdala, which interdigitate with each other and with elements of the magnocellular forebrain system. However, our conceptualization of the magnocellular basal forebrain also has evolved, in the sense that we now recognize that this collection of large neurons is neither exclusively cholinergic nor restricted in its distribution to a discrete basal nucleus of Meynert.

Magnocellular cholinergic and more recently described GABAergic and glutamatergic neurons contribute to an extensive collection of large, aspiny basal forebrain neurons that project to cortex and thalamus.[2] Most vividly distinguished in forebrain with cholinergic markers, such as antibodies against choline acetyltransferase or vesicular acetycholine transporters (Usdin, 1995; Arvidsson, 1997), this collection of neurons intertwines through and among several basal forebrain structures in serpentine fashion. In its most rostral extent, the basal forebrain magnocellular system occupies the medial septum-diagonal band complex. Caudal to this, it invades the subcommissural region, including the ventral pallidum (Grove et al., 1986; Záborszky and Cullinan, 1992). Further caudally, it is present in the sublenticular region, extending dorsally to occupy parts of the globus pallidus and fiber tracts related to it, including the internal and external medullary laminae and internal capsule (Grove et al., 1986; Henderson, 1997). At its most caudal limit, the basal forebrain magnocellular system occupies the anterior amygdaloid area (Gastard et al., 2002). Any impression one might have that these magnocellular neurons are scattered randomly through the forebrain is dispelled by the observation that the twists and turns of the elongated constellation they form are constant in different brains. This fact is emphasized in Mesulam's (1983b) numerical nomenclature describing the magnocellular corticopetal cholinergic neurons (see also Fig. 5.5). According to Mesulam's scheme, Ch1, Ch2, and Ch3 refer, respectively, to cholinergic neurons in the medial septum and vertical and horizontal limbs of diagonal band. Ch4 neurons include those in the subcommissural and sublenticular regions, and Ch5 and Ch6 neurons are located in

---

[2]Relevant primary literature includes Jones et al. (1976); Johnston et al. (1979); Lehman et al., (1980); Bigl et al. (1982); McKinney et al. (1983); Mesulam et al. (1983a); Woolf et al. (1983, 1984, 1986); Manns et al. (2001); Pearson et al. (1983); Rye et al. (1984); Saper (1984); Gritti et al. (1993, 1997); Sarter and Bruno (2002); Záborszky et al. (1986, 1991, 1999).

the pedunculopontine and laterodorsal tegmental nuclei, which are located in the mesopontine tegmentum.

Several investigators have observed a conspicuous connectional property of basal forebrain corticopetal neurons—that is, with very broad topography, they project back to the cortical areas that innervate the parts of basal forebrain in which they are located (Mesulam, 1983; Saper, 1984; Záborszky et al., 1991, 1999). Thus, cholinergic neurons in the septum-diagonal band continuum project to hippocampus, from which their major cortical innervation arises. Those in the horizontal limb of the diagonal band project to the olfactory system, from which they receive substantial projections. Frontal cortex projects robustly to parts of the Ch4 group harboring corticopetal neurons, which, in turn, project most strongly to frontal cortex. Caudal parts of the Ch4 located in the most posterior aspects of the horizontal limb of the diagonal band, posterolateral part of the globus pallidus and subjacent sublenticular region, receive strong projections from and project more prominently to parietal and occipital cortex (Saper, 1984). Finally, smallish cholinergic neurons that populate the caudal part of the sublenticular region and anterior amygdaloid area are reciprocally connected with the amygdalo-pyriform transition cortex.

3.7 SUMMARY

Unraveling the functional neuroanatomy of the basal forebrain has been a significant and noteworthy accomplishment of neuroscience research during the last three decades. The anatomical advances have exposed some fundamental organizational principles, which are of immediate relevance for the study of emotional and motivated behavioral expression and for understanding the pathophysiology of major neuropsychiatric disorders. The ventral striatopallidal system and the extended amygdala, which were introduced in this chapter, are intertwined in a complicated fashion with components of the basal nucleus of Meynert in the basal forebrain. The importance of these anatomical systems becomes apparent when one realizes that they receive some of their most salient projections from the orbitomedial and cingulate prefrontal cortices and insula, as well as from medial temporal lobe structures, including the basolateral amygdala, the hippocampal formation, and neighboring cortical regions. These cortical areas, all of which are integral parts of the greater limbic lobe, are increasingly implicated as the sites of primary pathology in major neuropsychiatric disorders, including temporal lobe epilepsy, schizophrenia, mood disorders, obsessive-compulsive disorders, drug addiction, and Alzheimer's disease. In the next chapter, we will discuss the greater limbic lobe structures and their connections.

4

THE GREATER LIMBIC LOBE

4.1 LIMITS, TOPOGRAPHY, AND RELATED CONCEPTS

As mentioned in previous chapters, Broca (1878), in a monumental effort entitled "Le Grand Lobe Limbique," described the limbic lobe in numerous mammals. He drew attention to the ring or limbus of each hemisphere that encircles the brainstem and forms its medialmost edge or border. Attaching the term *lobe* was significant, since it gave this cortex status as a separate entity like other lobular subdivisions of the cortex that had arisen in earlier years of the 19th century and during the 18th century. The cingulate gyrus dorsally and the parahippocampal gyrus ventrally are the major parts of the limbic lobe, but they are bridged together by many smaller areas of cortex, forming in total the complete cortical limbus.

Broca's deductions were purely topographical, formed without the aid of histological staining and microscopic analysis. Thus, when later-19th- and 20th-century anatomists interested in cortical cytoarchitecture began studying the cortex with differential staining and microscopy, careful scrutiny was given to the limbic lobe (Campbell 1905; Brodmann 1909). A bewildering set of descriptive terms emerged with time, and many that are still used today, necessitating comments on the limbic lobe, which go beyond topography alone.

In accordance with many previous authors (Yakovlev, 1972; Morgane et al., 1982; MacLean, 1990; Mesulam, 2000), we regard the limbic lobe as composed of the olfactory allocortex, hippocampal allocortex, and transitional cortical areas that intervene between these and the larger isocortices (Fig. 4.1). The transitional areas are numerous, form the bulk of the limbic lobe, and depart in terms of structure in one or more ways from the isocortex. For example, they may have a cortical layer absent, sparsely represented or exaggerated, and cellular sizes and density in cortical layers may differ substantially from the variable, but generally more uniform, isocortices. Finally, some limbic lobe areas and, interestingly, those related to memory, contain cell types and cellular aggregates in their superficial layers never found in the isocortex (Van Hoesen and Solodkin, 1993; Insausti et al., 1995; Solodkin and Van Hoesen, 1996). Limbic lobe cortical areas form the cingulate and parahippocampal gyri as well as the bridging caudal orbitofrontal, medial frontal, temporopolar, anteroventral insular, and retrosplenial cortices.

Thus, limbic lobe areas do not meet the rigid structural rules used to define isocortex in terms of cytoarchitecture, and this holds them together as a group, even though their structure may vary. Along with topography, these seemingly simple cytoarchitectural criteria help validate the concept of the limbic lobe. Innumerable neurological observations regarding anoxia, glucose utilization, chemistry, and epileptogenicity, as well as neuropathological observations, including neuro-oncology (Filimonoff, 1947; Yakovlev, 1959; Yasargil, 1994),

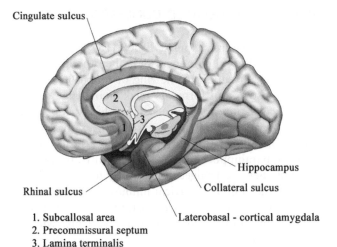

FIGURE 4.1 Schematic drawing of the medial surface of the human brain depicting the limbic lobe. The cortical areas of the parahippocampal gyrus have been removed to show the location of the underlying laterobasal-cortical amygdala and hippocampus. Reprinted from Heimer and Van Hoesen (2006) with permission. (Artwork provided by Medical and Scientific Illustration, Crozet, VA, USA.)

further buttress the segregation of limbic lobe cortex from the isocortex. In this context, it is important to revisit Vogt's theory of pathoklise (Vogt and Vogt, 1922; Vogt, 1925), which, to quote the late Pierre Gloor (1997), states that "certain physiochemical properties of nerve cells that share common morphological characteristics, and often constitute cytoarchitectonically definable areas, confer upon them specific susceptibilities to a variety of pathogenic agents." While much debated for a good part of the 20th century, pathoklise is now generally accepted and supported generously and unequivocally by chemical neuroanatomy. Differential neuropil- and phenotype-specific cellular staining methods, more often than not, reveal the uniqueness of one or more limbic lobe areas and are now standard procedures for the assessment of cortex.

Cortical parts of the limbic lobe, indeed, are disproportionally targeted by neurofibrillary tangles in Alzheimer's disease, and there is good evidence that this degenerative disease begins in selective nonisocortical parts of the limbic lobe (Hyman et al., 1984; Arnold et al., 1991; Braak and Braak, 1991; Van Hoesen, 2002). Considering the interest on altered cortical development in schizophrenia (Jakob and Beckman, 1986; Murray and Lewis, 1987; Weinberger, 1987; Bloom, 1993; Arnold and Trojanowski, 1996), the limbic lobe concept promoted here likely has special relevance to this debilitating disease, as well as many other psychiatric illnesses. Innumerable neuroimaging investigations link limbic lobe mechanisms to reward, motivation, feelings, emotion, and addiction.

4.2 THE GREATER LIMBIC LOBE

In addition to the nonisocortical areas just mentioned, we favor the inclusion of the laterobasal-cortical complex of the amygdala in the limbic lobe. Parts of it are cortex, and the various laterobasal nuclei contain cortical-like neurons aggregated into nuclei segregated by the intrinsic white matter of the amygdala. Moreover, chemicoarchitectonic characteristics and connectional patterns (Carlsen and Heimer, 1988; Heimer et al., 1999) support this viewpoint. The inclusion of the laterobasal amygdaloid complex in the limbic lobe is supported further by developmental investigations (Swanson and Petrovich, 1998; Puelles, 2001), which indicate that the laterobasal-cortical amygdala develops in association with the nearby populations of neuronal precursors that ultimately form the cortical mantle. By exclusion, and a host of other reasons discussed in Chapter 3, the remaining amygdala, the centromedial complex, belongs to the extended amygdala. Thus, the inclusion of the laterobasal cortical amygdaloid complex with the limbic lobe stems from a double disassociation inextricably linked to the compelling evidence that the centromedial complex stands apart and belongs to the extended amygdala.

There is an important caveat to be considered in regard to including major parts of the amygdala in the limbic lobe. Some in the past have included the entire amygdala in the limbic lobe, promoting what has been called "corticoid" histological features as well as its location within the parahippocampal region deep to the entorhinal and olfactory cortices. We acknowledge these features but point out the more compelling developmental, histological, and differential connectional data (Alheid et al., 1995; Heimer et al., 1997; de Olmos and Heimer, 1999; Alheid, 2003). Segregation of the limbic lobe and extended amygdala also reinforces what was obvious to classic anatomists and, more recently, physiologists—for example, Kapp et al. (1984)—that the amygdala is not a uniform functional/anatomical unit. The boundary between the laterobasal-cortical complex and centromedial complex is a critical line to draw both functionally (Davis and Whalen, 2001) and anatomically, as was emphasized by the neuroanatomist Brockhaus (1938) nearly seven decades ago for the human amygdala (see also Chapter 3). Further translation and discussion of this largely ignored and/or forgotten distinction can be found in Heimer et al. (1999).

It is not always appreciated that a sizable portion of the insular cortex located in the depths of the lateral fissure belongs to the limbic lobe and is a key bridging area of nonisocortex sandwiched between the temporal pole and the posterior orbitofrontal cortices. As shown in Fig. 4.2, this area corresponds

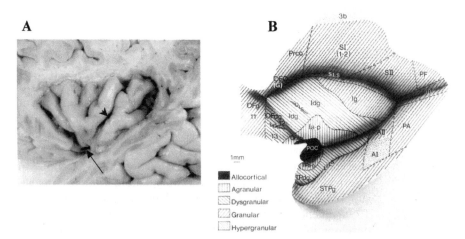

FIGURE 4.2 Dissection of the human insula in A. following removal of the frontal, parietal, and temporal opercular parts of the cortex. The arrowhead denotes the insular central sulcus, which separates the anterior short gyri from the longer posterior gyri. The black arrow denotes the location of the insula threshold or limen insulae. B denotes the various cytoarchitectural subdivisions of the insular cortex. Note the location of primary olfactory allocortex (POC) forming a semicircle around the limen insular and the agranular and dysgranular cortical fields. These components of the limbic lobe are bridging areas between the frontal and temporal lobes. (B. From Mesulam and Mufson, 1985. Reprinted from Heimer Van Hoesen (2006) with permission.)

to the anterior and ventral parts of the insular cortex, where cytoarchitectural analysis reveals a nonisocortical field (Mesulam and Mufson, 1985; Bonthius and Van Hoesen, 2006). Connectional studies link the anteroventral insular cortex powerfully to other frontal and temporal parts of the limbic lobe (Mesulam and Mufson, 1985). The direct continuity of this field with the cortex of the temporal pole and orbitofrontal cortex validates its inclusion as a bridging area in the continuous ring of cortex that forms the limbic lobe (Figs. 4.2 and 4.3). This area with long known visceral functional correlates, has been implicated in recent imaging investigations dealing with feelings and addictive urges (Naqui et al., 2007).

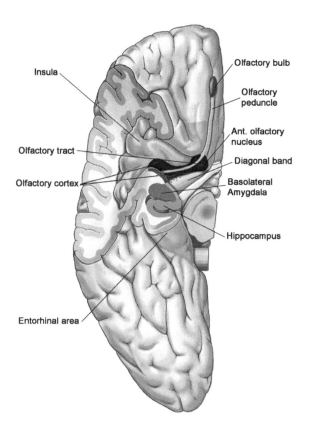

FIGURE 4.3 Ventral view of the human brain with the limbic lobe depicted in four colors. Darker green depicts the olfactory bulb and the location of the frontal and temporal primary olfactory allocortex. Lighter green depicts orbitofrontal, insular, and parahippocampal agranular, and dysgranular and periallocortical regions also belonging to the limbic lobe. Frontal opercular and temporal polar areas are dissected away to appreciate these and the location of the amygdala (orange) and hippocampus (violet). Reprinted from Heimer and Van Hoesen (2006) with permission. (Artwork provided by Medical and Scientific Illustration, Crozet, VA, USA.) See color plate.

CLINICAL BOX 6

The Insula

It increasingly comes to light that the role of the insula in the integration of normal sensations and motor functions and abnormal expression of the same in neuropsychiatric disorders has been underestimated. In seizures originating in the insula, a variety of visceral sensations, including gustatory and olfactory, may occur. Such seizures are associated with epigastric gurgling, rising epigastric sensations, nausea, and even vomiting. About 50 percent of patients with temporal lobe epilepsy show spontaneous spikes or spike-waves in the insula (Silfenius et al., 1964) and, on PET imaging, insular hypometabolism and benzodiazepine receptor loss have been reported (Bouilleret et al., 2002). The insula is also involved in the processing of pain (Peyron et al., 2004). After left anterior insular infarcts, speech impairments with slow initiation and poor fluency have been noted, and imaging studies reveal anterior insula activation with word generation tasks (McCarthy et al., 1993).

With regard to neuropsychiatric disorders, brain imaging studies have shown the anterior insula to be activated in a spectrum of anxiety disorders (Rauch et al., 1997), including Gilles de la Tourette's syndrome, obsessive compulsive disorder, and autism. It is involved in addictive behaviors, being activated by salient visual cues (Shelley and Trimble, 2004). The insula has been found to be smaller in schizophrenic patients than controls (Crespo-Facorro et al., 2000), and deficits in emotional expression have been noted in patients with insula lesions. Loss of insula tissue is also seen in frontotemporal dementia, Alzheimer's disease, and Lewy body dementia.

4.3 THE NONISOCORTICAL CHARACTER OF THE LIMBIC LOBE

We recognize the hippocampal formation and olfactory cortices as key elements of the limbic lobe (Figs. 4.3 and 4.4). These allocortical areas contain only two to three cortical layers, but they are composed of cortical neurons with distinct and polarized apical dendrites oriented toward the pia. Their neurons contribute layers, and/or are in continuity with certain layers that form multilayer transitional cortices sandwiched between the allocortices and the isocortices. For example, a peri-olfactory cortex extends along the olfactory allocortex dorsally into the insula, forming its agranular nonisocortical division. A similar cortex is present anteriorly along the frontal lobe olfactory cortex in the posterior orbitofrontal region. The perirhinal cortex (Brodmann's area 35) is also a peri-olfactory cortex, but it extends posteriorly along the entorhinal cortex, which forms the largest cortical field of the anterior paraphippocampal gyrus, partially covering the rolled up hippocampal allocortex. This

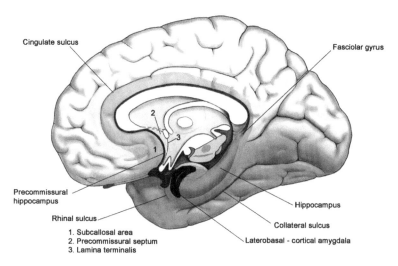

Cingulate sulcus

Fasciolar gyrus

Precommissural
hippocampus

Rhinal sulcus

Hippocampus

Collateral sulcus

Laterobasal - cortical amygdala

1. Subcallosal area
2. Precommissural septum
3. Lamina terminalis

FIGURE 4.4 Medial views of the human brain depicting the components of the limbic lobe in colors. Note the inner ring of allocortices in dark green, light purple, and dark purple, and the outer ring in light green, denoting the agranular, dysgranular, and dyslaminate cortices. Reprinted from Heimer and Van Hoesen (2006) with permission. (Artwork provided by Medical and Scientific Illustration, Crozet, VA, USA.) See color plate.

area is not shown in Fig. 4.4 so that the hippocampus can be visualized, but it is shown in Fig. 4.5. forming Brodmann's area 28. Also note in Fig. 4.5 that the perirhinal cortex (area 35) is depicted as a narrow, slit-like zone lateral to area 28, the entorhinal cortex. That the perirhinal cortex is invariably shown this way in cytoarchitectural maps of the human brain creates an important misconception regarding its size. In reality, this important area of the limbic lobe forms the major part of the medial bank of the collateral sulcus (Van Hoesen, 2002). This sulcus is always deep anteriorly, belying a buried but rather sizeable perirhinal cortex. Two-dimensional maps do not represent its size accurately.

The peri-olfactory nonisocortical areas just discussed above form an agranular component of the limbic lobe adjacent to the olfactory allocortex. The term *agranular*, incidentally, means simply that layer IV (the granular layer of the isocortex) is absent. Technically, however, another nonisocortical limbic lobe cortical area surrounds this initial peri-olfactory cortex before the isocortex is present. This outermost limbic lobe nonisocortical region is referred to as dysgranular, meaning simply, that a deficient layer IV is present but vaguely discernable under the microscope. Thus, layer IV of the dysgranular cortex is incipient—that is, seen sometimes but not always clearly or decisively. This dysgranular cortex forms large parts of the limbic lobe, including the dysgranular insular cortex, the posterior parahippocampal cortex, parts of the

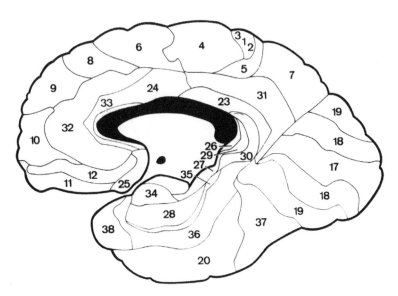

FIGURE 4.5 A map of the medial surface of the human brain depicting the localization of cortical cytoarchitectural fields after K. Brodmann, Vergleichende Lokalisationslehre der Grosshirnrincle, Barth, Leipzig (1909). (Reprinted from Zilles, 1990, with permission.)

posterior cingulate cortex, the medial cortex of the temporal pole, the posterior orbital cortex, and the medial frontal cortex. Together, the peri-olfactory agranular cortex and the dysgranular nonisocortex form the largest part of the limbic lobe.

The cingulate gyrus, largely comprising Brodmann's cortical areas 23, 24, and 25 (Fig. 4.5), is a sizable part of the limbic lobe that lies adjacent to the thin supracallosal part of the hippocampus known as the hippocampal rudiment (a.k.a., indusium grisium, shown in deep violet in Fig. 4.4). These peri-hippocampal areas are dominated by multiple pyramidal cell types exhibiting indistinct lamination—hence, they are referred to as *dyslaminate* (Vogt et al., 2005). Area 24, a good example, is sandwiched between the hippocampal rudiment and frontal lobe motor cortex (Brodmann's areas 4 and 6). The preponderance of pyramidal neurons in area 24 attests to its probable hippocampal lineage and continuity with the motor cortices (Sanides, 1969; Braak, 1976; see also Clinical Box 7).

To reiterate, much of the limbic lobe can be thought of as comprising belts of transitional cortex aligned adjacent to a core of olfactory and hippocampal allocortex. These are structurally agranular, meaning layer IV is missing; dysgranular, meaning layer IV is incipient; or dyslaminate, meaning that cortical

CLINICAL BOX 7

Cingulate Gyrus

The cingulate gyrus is an extensive structure in the human brain. In the past, the cytoarchitectural differences between anterior agranular and posterior granular parts of the cingulate region were emphasized, but more recently, progressing from anterior to posterior, at least four functional divisions are described, including viscero-motor, cognitive-effector, motor, and sensory processing regions (Mega and Cummings, 1997). From a neuropsychiatric point of view most attention has been paid to the anterior part of the cingulate gyrus, which is associated with disorders having much to do with motivation and movement, consistent with the extensive dorsal and ventral striatal connections of this part of the structure.

In animal experiments, stimulation of the cingulate cortex produces vocalizations (Jurgens and Ploog, 1970) and altered autonomic activity, including changes in respiration and heart rate, sexual arousal, and oral behaviors. Epilepsy originating in the cingulate gyrus is characterized by brief episodes of altered consciousness, vocalizations, rapid onset of motor activity (axial flexion and limb extension), and gestural automatisms (Devinsky et al., 1995). The human cingulate gyrus also has been shown to be nociceptive, probably related to its thalamic afferents. A wide variety of affective responses, including fear, euphoria, depression, and aggression, have been elicited upon stimulation of the cingulate gyrus in patients, as have behaviors characterized by disinhibition, hypersexuality, tic-like movements, and obsessive and/or compulsive activity. Lesions of the anterior cingulate gyrus produce one of the frontal lobe syndromes (see Clinical Boxes 5, 15 and 16) characterized by emotional blunting and decreased motivation.

Reduced blood flow (Bench et al., 1992) and tissue volume (Drevets et al., 1997) have been reported in the anterior cingulate gyrus of depressed patients as compared to controls. Drevets et al. (1997) have in particular drawn attention to decreased activity in the subgenual prefrontal cortex (ventral to the corpus callosum, corresponding to a part of the anterior cingulate gyrus), especially in untreated familial depression. Subsequently, a focus of increased metabolism centered within the area of decrease identified by Drevets et al. (1997) and associated with depressed mood was identified specifically in subgenual area 25 (Mayberg et al., 1999). High-frequency deep brain stimulation targeting area 25 has been shown in early trials to improve the clinical picture of some patients with chronic, treatment-resistant depression (Mayberg et al., 2005).

In schizophrenia, reductions in the numbers of small neurons, especially in layer 2 (Benes et al., 1993) and reduced cerebral blood flow (Tamminga et al., 1992) in the anterior cingulate cortex have been reported. As further regards structural changes in schizophrenia, however, numerous additional studies were done leading to a panoply of reports of no differences, asymmetries, smaller sizes, treatment effects, and gender differences. The subgenual region, consistently involved with the circuitry of depression, is reported to be unaffected in schizophrenia (Kopelman et al., 2005).

The less studied posterior part of the cingulate gyrus, as compared to the anterior, appears to be less involved with motor functions, having more to do with visuospatial function, learning, and memory, the last linked to cingulate connections with the subiculum (Devinsky et al., 1995). Some studies have linked hypometabolism in the posterior cingulate to depression, and the area blends together topographically and connectionally with the precuneus, an area of parietal cortex now thought to be involved with consciousness (see Basic Science Box 5).

Cingulate lesions have been carried out, albeit with diminishing frequency in recent years, in an attempt to relieve symptoms in several psychiatric disorders. Anterior cingulotomy, involving the bilateral removal of the anterior 4 centimeters of the cingulate gyrus, was done in patients with obsessive-compulsive disorders and chronic anxious tension but was less effective in depression or schizophrenia. Cingulotomy was also used in attempts to control intractable pain.

pyramidal neurons, particularly in deeper locations, are not organized clearly into layers.

The remaining parts of the limbic lobe, not accounted for in the preceding, constitute the memory-related retrosplenial, presubicular, parasubicular, and entorhinal cortices (Rosene and Van Hoesen, 1987; Amaral and Insausti, 1990). The deeper layers of each of these cortical areas are in continuity with the hippocampus, and each has a cell free lamina dissecans intervening between the deep and superficial layers, which have a distinctly laminar organization. This forms, in total, an atypical but multilaminar cortex. It seems inappropriate to label these perihippocampal cortices, since they are not dyslaminate. However, it is tempting to view them as hybrids formed by the overriding of the olfactory and hippocampal allocortices. Unfortunately, one can defend this notion only with topography and some connections. Helpful neurochemical correlates are sparse, as are developmental patterns from the embryo and molecular markers from the early stages of neurogenesis.

These cortical areas comprise that part of the limbic lobe in which are found unique neurons and aggregations of neurons with no counterparts elsewhere in the cortex. While the deep layers of these cortices are in continuity with the hippocampus, their superficial layers are aberrant with unknown origin. Possibly, they are early embryonic survivors and/or derivatives of cell types once present in the early development of the cortex but largely altered before the end-stages of cortical neurogenesis. Because of these questions, it is necessary, for the time being, to retain the term *periallocortex* and place the limbic lobe retrosplenial, presubicular, parasubicular, and entorhinal cortices in this category. This creates a limbic lobe formed by allocortex, periallocortex, peri-olfactory, perihippocampal, and amygdaloid components. While still having some terminology baggage, it enables one to jettison confusing terms like *paralimbic*, *mesocortex*, *schizocortex*, and *proiscortex*, to mention only a few.

4.4 THE OLFACTORY SYSTEM OF THE LIMBIC LOBE

The impetus for Broca's comparative study of the limbic lobe is not known with certainty. However, Broca had a lifetime interest in comparative anatomy, its functional correlates, and anthropology. The relative variation of the olfactory apparatus among species surely caught his attention, and the insertion of the olfactory tracts into the cortex he named the limbic lobe would be a compelling reason for him to undertake further study. He attributed olfactory sensation to the limbic lobe, but as Schiller (1979) points out in his excellent biography of Broca, he was cautious and, perhaps, even skeptical that the cortex of the whole limbus subserved only this function. Schiller's translations of the original text and Broca's notes reveal a critical, and somewhat contemporary-sounding, synthesis that comes very close to attributing visceral and emotion-related functions, along with olfaction, to the limbic lobe. This, along with the conceptual juxtaposition of olfaction and "limbic" functions that has emerged since Broca's report, necessitates a brief review of the neuroanatomy that underlies the entry of this sensory modality into the telencephalon.

The notion that the entire telencephalon was olfactory in primitive vertebrates characterized comparative and evolutionary thinking since their earliest times. However, as noted in Chapter 2, this was laid to rest in the early 1970s when it was shown that only a modest part of primitive shark telencephalon receives direct olfactory bulb input (Ebbesson and Heimer, 1970). This was followed shortly by the demonstration that the visual system and others have a representation in the telencephalon in primitive vertebrates (Ebbesson and Schroeder, 1971; Cohen et al., 1973). Nonetheless, olfaction and its cortical representation are an important force in the development of the telencephalon in vertebrate evolution, as Herrick (1956) so eloquently stated: "This olfactory field at the anterior end of the brain is the dominant center of control of all behavior in these primitive vertebrates, and for this reason it was the seedbed for further structural differentiation as the patterns of behavior were stepped up from one integrative level to another. Here, the rudimentary cortex had its beginnings."

This "driving force," on the part of the olfactory cortex, to integrate with other sensory systems in the telencephalon, is reflected clearly in the structural organization of the limbic lobe. The olfactory cortex and the peri-olfactory cortex that surrounds it have a close relationship with cortical regions representing other modalities, both in the caudal orbitofrontal cortex (Price et al., 1996) and in the neighboring insula (Figs. 4.3 and 4.4). Taste and somatic sensation (including visceral sensation) are also related to these areas. It is also relevant that the perirhinal cortex (Brodmann's area 35), another component of the peri-olfactory cortex, extends posteriorly all the way to the retrocalcarine region, where it abuts and/or forms area prostriata, the forerunner to the primary visual cortex (Morecraft et al., 2000; Ding et al., 2004). The relationship of these exteroceptive and interoceptive sensory systems to the olfactory system is not always appreciated. While they may represent less sophisticated visceral-oriented levels of sensory processing and sensory interactions, these are often the trigger zones for feelings and emotions, and it is noteworthy that the relevant areas are either within or closely adjacent to the limbic lobe. Before ending this discussion, it is useful to review the projections of the olfactory bulb in the human brain and in other higher vertebrates.

The target areas for olfactory bulb projections in the human are shown in Fig. 4.6. The depiction of these is based on normal anatomical studies in the human, extrapolations from experimental studies in monkey (Meyer and Allison, 1949; Heimer et al., 1977; Turner et al., 1978; Carmichael et al., 1994; Heimer et al., 1999), and an axon degeneration study in the human brain (Allison, 1954). Many olfactory bulb projections end in the anterior olfactory nucleus (retrobulbar area) where the olfactory peduncle first attaches to the orbital surface. Olfactory bulb fibers that don't end here form the lateral olfactory tract and end in the frontal and temporal primary olfactory cortex (piriform cortex) and the anterior perforated substance. The latter contains the laminated olfactory tubercle, a structure within which the striatum reaches the ventral surface of the brain (see Chapters 2 and 3). Although a medial olfactory tract was recognized by Broca (1878) and has since then been illustrated in nearly all textbooks, no experimental evidence has ever supported its existence (Heimer et al., 1977; Sakamoto et al., 1999; Price, 2004).

To reach the temporal piriform cortex, the lateral olfactory tract skirts across the limen insulae, which approximates the attachment of the frontal and temporal lobes. In humans this is an acute hairpin curve due to insertion of the temporal fossa beneath the bony orbit. Along this curve, olfactory bulb fibers fan out in several directions. One component (Fig. 4.6A–C) courses to the peri-olfactory agranular cortex of the insula. Another turns ventral and posteriorly along the anteriormost parts of the parahippocampal gyrus ending in the temporal piriform cortex, the anteriormost part of the perirhinal cortex and the anterior entorhinal cortex (Amaral et al., 1987; Insausti et al., 1995; Price, 2004). Olfactory bulb projections to the cortical amygdaloid structures in the human, finally, appear to reach the amygdalopiriform transition area and the anterior cortical amygdaloid nucleus.

In summary, olfactory input to the limbic lobe is substantial, reaching not only the primary olfactory cortex (frontal and temporal piriform cortex) but

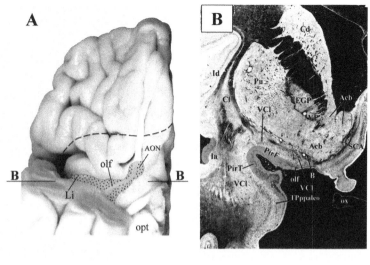

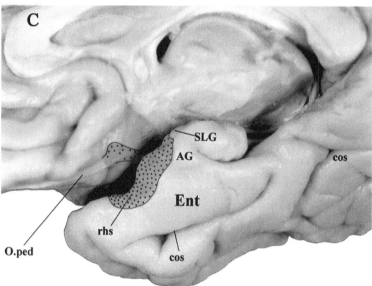

FIGURE 4.6 A–C: A is a photograph of the human orbital cortex following removal of the temporal pole. The interrupted line denotes the approximate anterior border of the limbic lobe, and the magenta-colored area denotes the projection zone of the olfactory bulb. The line labeled B-B denotes the line of cut for the coronal section shown in photograph B. As in A, the magenta color denotes the terminal zone for olfactory bulb projections. Photograph C shows the same, but it is mapped onto a ventromedial view of the human brain. Direct olfactory bulb projections extend beyond the temporal and frontal olfactory piriform allocortex and end in peri-olfactory agranular cortices in the insula, orbitofrontal, temporal polar, and parahippocampal regions. The latter includes area EO, the most anterior subdivision of the entorhinal periallocortex. Abbreviations: Acb—accumbens; AG—gyrus ambiens; AON—anterior olfactory nucleus; B—basal nucleus of Meynert; Cd—caudate nucleus; CL—claustrum; cos—collateral sulcus; EGP—globus pallidus; Ent—entorhinal cortex; Ia—agranular insula; Id—dysgranular insula; Li—limen insulae, olf—olfactory tract; O.ped—olfactory peduncle; opt—optic tract; ox—optic chiasm; PirE—frontal piriform cortex; Pirt—temporal piriform cortex; Pu—putamen; SCA—subcallosal gyrus; rhs—rhinal sulcus; SLG—semilunar gyrus; VCL—ventral claustrum. Note the abbreviation TPppaleo identifies Brodmann's area 35, the perirhinal cortex, a type of peri-olfactory cortex described in the text. Reprinted from Heimer and Van Hoesen (2006) with permission. See color plate.

CLINICAL BOX 8

Smell and Its Significance in the Clinical Setting

It is usually considered that the sense of smell is poorly developed in humans, in comparison, for example, to lower vertebrates, such as rats and mice. However, in view of its evolutionary significance, it seems very unlikely that the sense of smell is trivial in guiding human behavior.

The fragility of the fibers in the olfactory mucosa of the nose, their delicate passage through the cribriform plate, and the course of the olfactory tract along the orbital surface of the brain underly the vulnerability of this sensory modality to external trauma—for example, from head injury. This is compounded by the locations of structures receiving the central representations of smell, such as, for example, the primary olfactory cortex and olfactory tubercle, which lie vulnerable on the undersurface of the brain overlying the anterior perforated space.

Smell sensations may occur as part of an aura in temporal lobe epilepsy. Traditionally referred to as uncinate seizures (simple partial seizures in today's terminology), the seizure focus was thought to be in the uncus, overlying the amygdala. However, olfactory inputs also end in the anterior insula, which thus may be associated with the experience. These aurae therefore have some localizing value, but they are not of lateralizing significance.

The olfactory system is involved in several neuropsychiatric disorders. In depression, for example, the sense of smell can be diminished, as may be other sensory modalities, such as taste or touch. Diminished smell sensation also has been reported in Alzheimer's disease and Parkinson's disease, especially the Lewy body variant but not sufficiently reliably to be used in any diagnostic way (Hawkes, 2003). Diminished smell also may be observed early in the course of schizophrenia and may be associated with smaller perirhinal cortices as measured with MRI (Turetsky et al., 2003). Thus, disturbances of smell in such disorders may be associated with underlying neuroanatomical deficits, rather than being simply a manifestation of a psychosis or deteriorating intellect.

also the periolfactory cortex that surrounds it. From the latter it is positioned to reach substantial parts of the distal association cortices, other parts of the limbic lobe, including the hippocampus, and the frontal lobe via corticocortical connections (Van Hoesen, 1982; see also Clinical Box 8).

4.5 OTHER SENSORY INPUT TO THE LIMBIC LOBE

When Papez (1937) formulated his famous circuit for emotion, involving major portions of the limbic lobe, little was known about sensory input to the circuit other than olfaction. MacLean (1949, 1952) had similar problems, thus leaving the emotive process without input from the environment or periphery to trigger or grade its expression. It was two decades later before experimental neuroanatomical studies began filling this problematic void. The classic reports of Pandya and Kuypers (1969) and Jones and Powell (1970) on corticocortical connections in the monkey were key in reversing this. Using more sensitive experimental methodology, these investigators asked the basic question, what are the targets of cortical sensory information leaving the primary sensory cortex? After successively tracing series of corticocortical connections from the primary sensory cortex to proximal, and then more distal, sensory association cortices, three critical targets of the latter were observed. One system of axons is directed to the frontal association cortices, a second is directed to multimodal areas of cortex, and a third to targets within the limbic lobe (Fig. 4.7). Further observations indicated that frontal and multimodal association cortices also project to parts of the limbic lobe (Van Hoesen et al., 1972; Van Hoesen, 1982; Witter et al., 1989; Witter, 1993; Suzuki and Amaral, 1994). Of great interest was the fact that these association inputs to the limbic lobe are both sizeable and distributed selectively. Together, these findings argued for widespread functional diversity within the limbic lobe, since limbic lobe target areas received diverse components of sensory associative output.

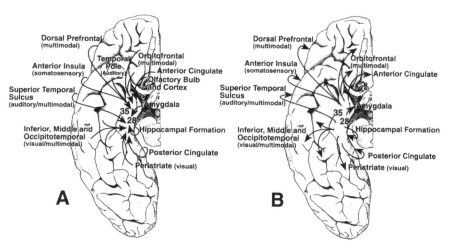

FIGURE 4.7 A–B Input (A) and output (B) connections of the human entorhinal (area 28) and perirhinal (area 35) cortices shown on ventral views of the hemisphere. These are deduced from experimental research results in the monkey and drawn from several laboratories, cited in the text. Note that the major sensory modalities are represented, as well as multimodal input. The input connections are known to be associated with episodic memories and the output connections with their consolidation in the association cortices. The neurofibrillary tangles of Alzheimer's disease damage areas 28 and 35 heavily compromising these key input and output hippocampal areas of the limbic lobe. Panel A is reprinted from Heimer and Van Hoesen, 2006, with permission.

Although many volumes could be written on the neuroanatomical, neurophysiological, behavioral neurology, and neuropsychiatric research launched by these decisive findings, a brief outline of cortical association connections is warranted here with regard to the limbic lobe (Pandya and Yeterian, 1985). Information related to the visual modality is dispersed to temporal association cortices before relay to the perirhinal and posterior parahippocampal cortices and basolateral amygdala. All agree that these pathways are critical for many forms of behavior including fear, face recognition, declarative memory, and object recognition. The hippocampus and amygdala are key players in the limbic lobe associated with these behaviors, as is known from both the clinic and the experimental behavioral laboratory.

Auditory sensory information relays extensively to the auditory association cortices that form the superior temporal gyrus. These course anteriorly to reach the limbic lobe cortex that forms the temporal pole, with additional projections to the basolateral amygdala, entorhinal cortex, posterior orbital cortex, and insular and posterior parahippocampal gyrus, again all components of the limbic lobe. As with the visual cortices, the auditory association cortices send axons to the frontal lobe and to multimodal areas located in the depths and banks of the superior temporal sulcus. Sound localization, auditory memory mechanisms, species-specific vocalizations, and the perception of threat-related stimuli are all thought to be associated with these pathways, as well as, of course, linguistically related processes in humans.

Somatosensory information from the postcentral gyrus courses extensively to motor and premotor areas within and anterior to the central sulcus but also somatosensory association areas that form the superior and inferior parietal lobules. The latter, in particular, has extensive projections to the frontal association cortices (Petrides and Pandya, 1984) and multimodal association cortical areas in the superior temporal sulcus. Some of the latter contribute to the formation of truly trimodal association areas with somatosensory, auditory, and visual input to a single locus. Other areas are known to be bimodal. Importantly, however, parietal association projections course to many limbic lobe areas, including the presubiculum, posterior parahippocampal cortex, the dysgranular insular cortex, and much of the cingulate gyrus (Vogt and Pandya, 1987; Van Hoesen et al., 1993). Association cortex on the medial surface of the hemisphere that forms the precuneus is reciprocally connected with much of the cortex of the inferior parietal lobule and has similar limbic lobe, mutimodal, and prefrontal connections. Since the precuneus is continuous with the cortex of the posterior cingulate gyrus, its connections with this part of the limbic lobe are especially rich (see Basic Science Box 5). Body representation and localization, pain, spatial localization, response selection, temperature sensation, and visceral sensory perception are some of the many somatic related processes carried by the neural systems just discussed. Their entry into limbic lobe involves multiple targets within it from many levels of integrative processing within the parietal lobe dealing with the body and space. A recent review (Cavanna and Trimble, 2006) of the functional correlates of the little-known precuneus concludes that this cortex may play a special role in mental imagery related to the self.

In summary, early investigators were preoccupied and, probably, overimpressed with olfactory sensory input to the limbic lobe. Even in humans, this is direct and significant, but in the last four decades it has been learned that sensory input from all modalities influence the limbic lobe, and olfaction is simply one of these. The quantity and directness of this input in primates are impressive. Furthermore, sensory information also converges into multimodal association areas of the frontal and temporal lobes, and these then send axons to the limbic lobe, providing it with input with additional levels of complexity and integration.

Before leaving this topic to consider limbic lobe outputs, it is critically important to appreciate the functional implications of sensory input to the

BASIC SCIENCE BOX 5

Precuneus

The cortical areas that form the lateral parts of the parietal lobe are the great and well-known provinces of somatic sensation, body space, and environmental space. Visceral sensations are represented in the enfolded cortex of the parietal operculum and insula, including parts of insular agranular and dysgranular limbic lobe. The functional correlates of the medial parietal cortex forming the precuneus are significantly less known, and little has been available about its neuroanatomical correlates. A recent investigation (Parvizi et al., 2006) in the monkey has revealed that the medial parietal cortex is extensively interconnected with the cingulate gyrus of the limbic lobe and multimodal association cortices in the frontal, parietal, and temporal lobes. Numerous thalamic and other subcortical connections also point to a uniqueness of the precuneus and adjoining posterior cingulate gyrus. Functional investigations of the precuneus, largely with imaging methodology, have accrued with rapid pace in recent years, and many suggest that this cortex subserves a higher-order behavior relating to the conscious self-percept (Cavanna and Trimble, 2006). The precuneus is one of the sites in the brain where increased metabolic activity occurs while the brain is idling or the individual is self-reflecting. Further, activity in the precuneus decreases in states of diminished self-awareness, such as anaesthesia, dementia, sleep, or when attention is engaged in other tasks (Vogt and Laureys, 2005). The higher-order ability to cognitively conceptualize and project one's self into both the future and past may be a function of this cortex, supported neuroanatomically by its extensive association and limbic lobe connections, and output to basal forebrain.

limbic lobe. These connections are not a simple serial cascade of cortical association axons. Instead, extensive distributed processing occurs within the various sensory association cortices, and downstream projections from them to distal and multimodal association areas and the limbic lobe comprise multiple parallel systems, likely laden with highly integrated information. Also, it is critical to keep in mind that the same sensory association areas involved in real-time information processing from the environment are also the historical repositories of past experiences. Thus, what the limbic lobe receives is a combination of real-time integrative sensory processing weighted against past experiences, memories of these, previous reactions to such stimuli, acquired personal and societal norms, and the state of the perceiver at that point in time. These design features of sensory-limbic lobe connections when operating normally provide, among many things, optimal circumstances for the detection of novelty and new learning, the detection of danger and/or threat and cues relevant to its avoidance, and the detection of the behaviors of others, including facial expressions, which are critical for human interactions. When the real-time and historical record of sensory input to the limbic lobe is altered, the doorway is opened for abnormal behavior on all of these and, likely, to mental illness. The historical component of sensory information to the limbic lobe is key to understanding the often long-accrued developmental nature of emotional disorders (see Clinical Box 9).

4.6 LIMBIC LOBE OUTPUT

The projections of the limbic lobe to cortical and subcortical areas are extensive, and we will focus mainly on three of these: projections back to the association cortex (Figs. 4.7 and 4.8A and B), projections to a cortical motor area, and subcortical projections to the basal forebrain.

In general, it is accurate that the limbic lobe reciprocates all of the projections it receives from the cortical association areas, although a full experimental documentation of this is not complete, and some important exceptions occur. For example, the somato/visual association cortex of the parietal lobe

CLINICAL BOX 9

Limbic Lobe in a Clinical Context

In clinical neurology it is customary to ascribe certain signature syndromes to various lobes of the cerebral cortex. Thus, there are well-recognized parietal lobe syndromes, such as the apraxias and alexias and, of course, the disinhibitory

frontal lobe syndromes. However, few texts describe specific limbic lobe disorders, although there are several that would fall under this rubric. These have
been instrumental in helping reveal the functions of the limbic lobe in human
behavior, complementing studies of animals in the laboratory and brain imaging
studies in human volunteers.

Limbic encephalitis is a rare disorder, secondary to certain viral infections,
including, notably but rarely, rabies and, more commonly, herpes simplex (HSV-
1) infection. Following a prodromal phase of nonspecific symptoms, behavioral
changes occur following infections by HSV-1 that include bizarre, out-of-
character behavior, sometimes with hallucinations and epileptic seizures. The
EEG may reveal unilateral or bilateral changes in temporal leads. HSV-1 viral
infection leads to marked pathology in frontal and temporal areas, and the
chronic state is characterized by dense amnesia, irritability, aggression, distract-
ibility, and emotional blunting—in some ways reminiscent of the Klüver-Bucy
syndrome, one of the first limbic syndromes to be recognized. The Klüver-Bucy
syndrome, following bilateral destruction of the temporal lobes, including the
amygdala (Weiskrantz, 1956), is characterized by an increase in sexual activity,
a compulsive tendency to place objects in the mouth, decreased emotionality,
visual agnosia, and changes in eating behavior. These behavioral changes have
been seen in clinical settings, notably after head injury with damage to both
temporal lobes, and in some cases of dementia, especially Pick's variant. Similar
clinical signs may also be seen in Alzheimer's diesease, but the onset is much
earlier in Pick's disease (Cummings and Duchen, 1981). The visual defect in the
Klüver-Bucy syndrome is in all likelihood related to involvement of the temporal
neocortex rather than of the amygdaloid body. Furthermore, it is unlikely that
a specific part of the amygdala or the surrounding cortical regions relates to each
one of these complex behaviors. Rather, the various symptoms are likely to be
related to a more global defect, which makes it difficult to relate sensory information to past experience or to evaluate sensory stimuli in terms of their emotional and motivational significance.

A variant of limbic encephalitis is seen in some patients with carcinoma who
present with disturbances of affect, depression, anxiety, and sometimes hallucinations and/or seizures. Such symptoms are seen especially in patients with
bronchial oat cell carcinomas, and pathology is noted in the amygdala, hippocampus, fornix, and mammillary bodies. This pathology is thought to be secondary to an immunological process, but its tendency to target the limbic lobe
relatively selectively suggests that the neuronal constitution underlying this specific anatomical localization may involve some defining biological markers.
Other limbic lobe pathologies include temporal lobe neoplasms and epilepsies,
in which the pathology is often quite specific, with cell loss and gliosis in the
hippocampus and sometimes also the amygdala (see Clinical Boxes 11 and 12).
It has been known for a long time that there is some association between brain
tumors and schizophrenia, especially tumors of the amygdala-hippocampal
region (Malamud, 1975).

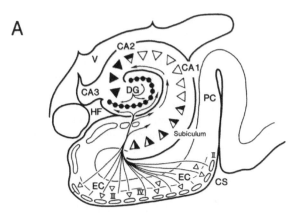

HIPPOCAMPAL CORTICAL INPUT

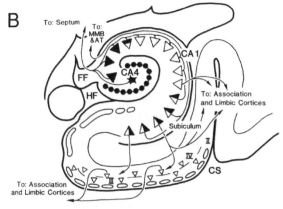

HIPPOCAMPAL CORTICAL OUTPUT

FIGURE 4.8   A–B: These illustrations depict the details of hippocampal formation input (A) from the entorhinal cortex and output (B) from its subicular and CA1 divisions. The former form the compact perforant pathway, while output pathways are generally more diffuse, using multiple hemispheric white matter pathways. Together, these link the hippocampal formation to the association cortices. Abbreviations: AT—anterior thalamus; CA 1–3—the major pyramidal cell subdivisions of the hippocampus; CS—collateral sulcus; DG—dentate gyrus; EC—entorhinal cortex; FF—fimbria-fornix; HF—hippocampal fissure; MMB—mammillary bodies; PC—perirhinal cortex. Reprinted from Van Hoesen, 1997, with permission.

projects to parts of the cingulate gyrus, and this part of the limbic lobe sends axons back to it (Fig. 4.9). Superior temporal gyrus auditory association areas project to the orbitofrontal, temporal polar, and posterior parahippocampal gyrus, all parts of the limbic lobe, and these areas project back to the superior temporal gyrus. A similar reciprocity seems to exist between prefrontal and multimodal association cortices and the limbic lobe. For example, the granular

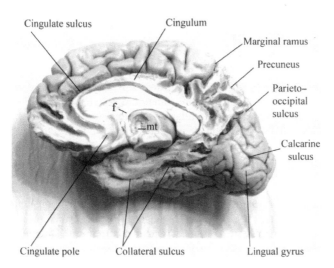

Cingulate sulcus

Cingulum

Marginal ramus

Precuneus

Parieto-occipital sulcus

Calcarine sulcus

f

mt

Cingulate pole

Collateral sulcus

Lingual gyrus

FIGURE 4.9 This photograph shows a dissection of the cingulum bundle in the human brain. Note, its course around the corpus callosum and into the parahippocampal region medial to the collateral sulcus. This long white matter links much of the limbic lobe together. However, also note its many side branches into the medial association cortex of the parietal lobe, known as the precuneus, and similar branches into the orbitomedial cortex of the frontal lobe. As described in the text, these association areas of the frontal and parietal lobes have a special connectional relationship with the limbic lobe, providing major input and receiving limbic lobe output. Reprinted from Heimer and Van Hoesen (2006) with permission.

prefrontal cortex sends axons to the posterior cingulate cortex, including its retrosplenial part, and receives back a strong reciprocal cingulate projection. The multimodal cortex of the superior temporal gyrus sends projections to limbic lobe areas, like the posterior cingulate cortex, and these areas reciprocate the projections.

Some important exceptions to this widespread reciprocity occur, however, and we will address two of them. It is known that the entorhinal periallocortex receives cortical afferents from many sources, the majority of which arise from other parts of the limbic lobe (Van Hoesen, 1982; Amaral et al., 1983; Insausti et al., 1987; Suzuki and Amaral, 1994). It in turn gives rise to the perforant pathway, which ends in hippocampal formation, linking this structure with the cortex (Figs. 4.7 and 4.8A and B). The hippocampal formation itself receives little, if any, direct cortical input except from the entorhinal and perirhinal cortices. However, its CA1 and subicular divisions send extensive projections back to cortical areas from which they do not receive direct projections (Rosene and Van Hoesen, 1987). These pathways, known to be involved in memory consolidation, are diffuse but extensive. Both the CA1 and subicular division of the hippocampal formation are damaged heavily in Alzheimer's disease (Hyman et al., 1984; Van Hoesen et al., 1991; see also Clinical Box 10).

CLINICAL BOX 10

Entorhinal Cortex

Brodmann's area 28 forms the entorhinal cortex, the anterior part of the parahippocampal gyrus, medial to the rhinal and collateral sulci. The smaller area 34 lies medial to area 28, is similar in structure, and forms the gyrus ambiens. In approximately 70 percent of human brains, the free edge of the tentorium cerebelli contacts, and grooves or notches, the anterior parahippocampal gyrus. The gyrus ambiens lies medial to the tentorial notch unprotected by dura in the tentorial aperature (Fig. A). With increased intracranial pressure in the supratentorial space, the gyrus ambiens can easily herniate, followed by the uncal hippocampus in severe instances. This pathology can compromise the third cranial nerve, alter the flow of CSF through the tentorial aperature, and compress the brainstem, necessitating immediate clinical attention. Traumatic head injury by accident or difficult childbirth can scar the entorhinal area along the rigid free edge of the tentorium cerebelli.

Unlike all other areas of the cerebral cortex, the entorhinal periallocortex of the limbic lobe can be localized without the aid of magnification, since its surface is covered by small wart-like elevations known as the *verrucae hippocampi* (Fig. A). These were described by the Swedish neuroanatomist Gustav Retzius (1896; Das Menschenhirns: Studies in der makroskopischen morphologie. Norstedt and Sohne, Stockholm), who commented on the similarity of their appearance to a viral skin disease and the skin of some amphibians. Each elevation

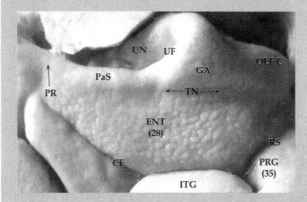

CBOX 10 FIGURE A Photograph of the ventromedial temporal lobe of the left hemisphere showing the anterior parahippocampal gyrus and its associated entorhinal cortex. Note the verrucae hippocampi that mark its surface. Abbreviations: CF—collateral fissure or sulcus; ENT (28)—entorhinal cortex (Brodmann's area 28); GA—gyrus ambiens; ITG—inferior temporal gyrus; OLFC—primary olfactory cortex; PaS—parasubiculum; PR—perirhinal cortex; PRG (35)—perirhinal gyrus (Brodmann's area 35); RS—rhinal sulcus; TN—tentorial notch; UF—uncal fissure; UN—uncus.

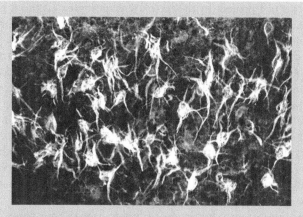

CBOX 10 FIGURE B Photomicrograph of thioflavin-S stained neurofibrillary tangles from layer II of the entorhinal cortex in Alzheimer's disease. These clusters of neurons underlie the verrucae hippocampi, which are flattened in this disorder.

marks the location of an island of large multipolar neurons that forms layer II of the entorhinal cortex (Solodkin and Van Hoesen, 1996). The interspaces between the islands are filled with small inhibitory neurons and/or undifferentiated immature neurons. The multipolar neurons are some of the first affected by neurofibrillary tangles in aging and Alzheimer's disease (Fig. B). They give rise to a major part of the perforant pathway which is a critical cortico-hippocampal linkage for episodic memory (Van Hoesen, 2002). Jakob and Beckman (1986) have noted their developmental abnormality in schizophrenia, and they, along with other entorhinal layers, are implicated in numerous neurological and psychiatric disorders (Arnold and Trojanowski, 1996).

Despite these many clinical correlates, precious little additional information is known about the verrucae hippocampi. Why do they form? Are they present in children as well as adults? What is their life history? Do immature interspace neurons replace or join multipolar island neuron populations? Are verrucae evenly distributed in the human population? Why are verrucae neurons first affected in aging and Alzheimer's disease? And do entorhinal developmental abnormalities contribute to the symptomatology of schizophrenia? These are but a handful of questions that could be formulated about the verrucae hippocampi that await further basic and clinical neuroscience research.

Another example of nonreciprocal limbic lobe output to the cortex involves the basolateral amygdala. These nuclei receive extensive limbic lobe and distal association cortex input (Herzog and Van Hoesen, 1976; Van Hoesen, 1981; Amaral et al., 1992). In many instances amygdalo-cortical output reciprocates these. However, in the visual system, the amygdala projects to the proximal peristriate association cortices, and even to primary visual cortex itself, both of which do not project to the basolateral complex (Amaral et al., 2003). In this sense, the amygdala can influence visual processing at its first cortical level of analysis and modulate the input it receives at all cortical levels: primary sensory, proximal association, and distal association. Given the amygdala's known role in such processes as fear conditioning and the perception of facial expressions, both of which could be critical to the well-being of the organism, it is tempting to believe that this limbic lobe output to the cortex has the element of safety, with millisecond timing built into it for self-protection (see Clinical Box 11). In summary, the preceding brief reviews and the discussion of sensory input to the limbic lobe in the prior section establishes the fact that the limbic lobe is powerfully interconnected within itself and with distal association and multimodal and prefrontal cortices. As recently as four decades ago, little of this was known.

Another important component of limbic lobe output concerns its influence on motor-related parts of the cerebral cortex. This is a more limited feature of its output that is focused on Brodmann's area 24, the major element of the dyslaminate limbic lobe cortex forming the anterior cingulate gyrus. Here, in the lower bank and fundus of the cingulate sulcus, complete motor representations can be found for the body with all parts represented (Braak, 1976; Biber et al., 1978). These areas are known as area M3 and M4 (Morecraft and Van Hoesen, 1992) or simply cingulate motor areas, and they give rise to corticobulbar and corticospinal axons (Dum and Strick, 1991; Morecraft et al., 2001). Moreover, they are reciprocally interconnected in terms of body part representation with both the supplementary motor cortex (M2) and the primary motor cortex (M1) in the precentral gyrus (Morecraft and Van Hoesen, 1992). Hence, an input to any one body part, can trigger activity in homologous parts of all of these multiple cortical representations. On the subject of emotional-motor interaction, it is significant that the anterior cingulate motor area M3 receives a massive input from all parts of the limbic lobe (Morecraft and Van Hoesen, 1998). An especially strong input from the basolateral complex of the amygdala specifically targets the face representation of M3 (Morecraft et. al., 2007). It is thought that this system of axons plays a role in emotional facial expressions, voluntary mimicry, and orofacial automatisms observed in temporal lobe seizures. In view of the presence of Lewy bodies in the basolateral complex of the amygdala in Parkinson's disease, one might also hypothesize that these limbic-motor connections account for the clinical sign of masked facies commonly observed in that disorder.

A third and especially massive output of the limbic lobe is directed to the basal forebrain, where all components receive limbic lobe input. For example, the subicular division of the hippocampal formation sends powerful projections via the fornix that end in medial portions of the accumbens and other medially located islands of striatal neurons. The deep layers of entorhinal cortex, which receive a large projection from the subicular division of the

CLINICAL BOX 11

Amygdala

In terms of interest to clinicians, the amygdala has until recently held second place to the hippocampus. The anatomical location and structure of the hippocampus have rendered it easier to study experimentally, and facilitated the comparison of animal data to damage found in certain clinical situations, especially temporal lobe epilepsy. In contrast, the boundaries of the amygdala are more complex, especially with the now accepted neuroanatomical extension of amygdala through the subpallidal region and into the bed nucleus of the stria terminalis, comprising part of the extended amygdala. Furthermore, in clinical practice there are few conditions that involve almost exclusively the amygdala. Nonetheless, experimental evidence links the amygdala to the modulation of emotion, in contrast to the association of the hippocampus with cognitive and episodic memory functions.

The clinical relevance of the amygdala was first appreciated following the description of the Klüver-Bucy syndrome (described in greater detail in Clinical Box 9), wherein bilateral removal of the amygdala in monkeys leads to taming and placidity, a tendency to explore the environment orally, an environmental distraction (hypermetamorphosis), and increased sexual activity.

The amygdala is subject to bilateral degeneration in the Urbach-Weithe disease. On the basis of clinical observations in these patients, and experiments in which people are imaged with fMRI while viewing faces of varying emotional valence, Adolphs et al. (1999) suggested that the amygdala is involved in the social appraisal of the emotional state of others, especially for negative emotions, but also is implicated in a broader spectrum of social attributions, related to value judgements such as trustworthiness, and other complex social emotions. Providing emotional valence to sensory stimuli, and via widespread reciprocal connections with neocortex and a variety of subcortical structures, it can set the level of sensitivity of the individual to incoming environmental events, attaching emotional color to the percepts. Further, amygdala is linked to the emotional tone of memory consolidation and restructuring. When the amygdala speaks, the entire brain listens.

There has been a lot of experimental evidence that the amygdala is part of a circuit related to aggressive behaviors. The initial observations of Klüver and Bucy (1937) were followed by many other studies showing that stimulation of structures such as the amygdala, the hypothalamus, or the diencephalic fornix elicited aggression, which could be inhibited by stimulation of the ipsilateral frontal cortex. In animal studies, amygdala integrity has been shown to be important for maintenance of position in a social hierarchy. Associations between amygdala and frontal pathology and aggression have been shown in patients with temporal lobe epilepsy, aggression being one behavior reliably reduced following temporal lobectomy as a treatment for intractable epilepsy (Trimble and Tebartz van Elst, 1999).

With such a key role in the limbic pantheon, the amygdala not surprisingly has been shown to be involved in the underlying pathology of several psychiatric disorders. For example, in psychiatric practice patients are often evaluated who have exhibited spontaneous outbursts of aggression, but in whom no evidence of an epileptic seizure has been reported. This picture may be contrary to the patient's personality characteristics and may be noted after minor head injuries. The aggression may be provoked by minimal provocation, or by the taking of a small amount of alcohol. This was referred to as episodic dyscontrol, but in DSM IV is called intermittent explosive disorder. Such patients often have abnormal EEGs especially with slow waves, sharp waves, or sometimes spikes from the temporal areas, and may respond to antiepileptic drugs. They should not however be diagnosed as having epilepsy, and should also be distinguished from patients with long-standing histories of personality disorder. On brain imaging some such patients will reveal cysts or even tumors of the amygdala.

Amygdala auras reveal the onset of an epileptic seizure, and include déjà vu phenomena, rising epigastric sensations, and feelings of panic. The latter can be easily be mistaken for panic disorder, although they tend not to be triggered by environmental circumstances, and tend to be of more sudden onset and offset and of shorter duration. Direct stimulation of the amygdala leads to anxiety feelings and fear. In epilepsy, an aura of fear is mainly lateralized to the right amygdala, as is that of déjà vu. Amygdala activation has been shown during symptom provocation in post-traumatic stress disorder (Rauch et al., 2000), panic disorder, and social phobia (Tillfors et al., 2001). In the affective disorders amygdala involvement is reflected by autonomic symptoms (anxiety, tension, agitation), but also with the patient's altered emotional evaluation of the environment, which is typically laden with negative affect, unpleasant memories, and pervading pessimism.

Several brain imaging studies cohere with a finding that amygdala metabolism is increased during major depression in the resting state, and decreases with treatment and clinical improvement (Drevets 2003). The amygdala of depressed patients exhibits increased responses to emotional stimuli compared with controls (Sheline et al., 2001), but data on the size of the amygdala in depression is variable. Of some interest however are repeated observations of increased amygdala volume, mainly bilaterally, in patients with bipolar disorder, especially in adults but not in adolescents (Altschuler et al., 1998; Strakowski et al., 1999). This has not been attributed to medication effects, and it is unclear if it relates to state or trait characteristics. However, it is concordant with findings of increased amygdala metabolism noted in bipolar disorder and elevated medial limbic activity reported in euthymic bipolar patients (Mah et al., 2007). In schizophrenia the amygdala is smaller in size than in matched controls in a population at risk for the condition and smaller again in established schizophrenia (Lawrie et al., 2003).

hippocampal formation, also send input to the accumbens that overlaps the zone of direct subicular projections within the accumbens (Sorenson and Witter, 1983). It is noteworthy that subicular output from the hippocampal formation, and deep layers of the entorhinal cortex, is known to play a role in memory consolidation, and it is tempting to believe that such outcomes are highly relevant to basal forebrain functions as well (see Clinical Box 12).

The basolateral amygdaloid complex is a major contributor of axons and input to the basal forebrain (Amaral et al., 1992; Fudge et al., 2002). These projections are diffuse in nature but stream into the ventral telencephalon to end in the accumbens (shell and core) and the collections of striatal neurons that characterize much of the ventral striatum. Importantly, basolateral amygdaloid projections terminate heavily among the islands of large neurons that form the nucleus basalis of Meynert—that is, among the so-called magnocellular cholinergic, GABAergic, and glutaminergic elements of the basal forebrain. Since these project widely to the cerebral cortex, providing its major cholinergic input, it is clear that the limbic lobe exercises a major influence on this excitatory neurotransmitter system (Kievet and Kuypers, 1975; Mesulam and Van Hoesen, 1976). We have already mentioned entorhinal cortex input to the basal forebrain, but this is one of many cortical inputs. Indeed, the perirhinal, temporal polar, insular, orbitofrontal, medial frontal, and anterior cingulate limbic lobe cortices all send strong quantities of axons to the basal forebrain (Van Hoesen et al., 1976; Mesulam and Mufson, 1984; Haber et al., 1995; Chikama et al., 1997; Ferry et al., 2000). These, collectively along with basolateral amygdala projections, constitute a truly massive system of afferent input. While it is quantitatively difficult to document such a statement, it is certainly tempting to hypothesize that the major output of the limbic lobe is directed toward the basal forebrain. It is useful in this regard to recall that much of this input has to be viewed as the end station of cortical integrative processing. It entails distributed and parallel series of corticocortical connections carrying both a record of what is being perceived—that is, real-time information—influenced by the nonlabile records of past history stored or present in the association cortices.

4.7 CONCLUDING REMARKS

In many species without gyrencephalic cortices, the limbic lobe can form an innocent-appearing band of cortex around the medialmost edge of the hemisphere. Even in species like nonhuman primates and humans with prominent gyri and sulci, the major components of this lobe are apparent with cursory inspection, since they are largely uninterrupted by sulci. A diversion into the lateral fissure can be confusing until the anterior insular and posterior orbitofrontal cortex are appreciated as bridging areas. This superficially simple

CLINICAL BOX 12

Hippocampus

The hippocampus has long interested clinicians and has been associated with psychoses for nearly two hundred years, following observations by Bouchet and Cazauvieilh (1825) of hippocampal pathology in patients with epilepsy and psychoses.

The hippocampus receives cortical input primarily from the nearby entorhinal and perirhinal cortices, which themselves receive substantial diverse inputs from many areas of cortex. Subcortical input reaches it via the fornix and diffuse systems in its vicinity, which link the hippocampus to the amygdala and other basal forebrain structures. Hippocampal output is both compact and diffuse. The former uses the fornix and links hippocampus to the septum, accumbens, anterior thalamus, and mamillary body. The latter diffuse hippocampal output systems use cortical white matter and provide hippocampal output directly to nearly all of the limbic lobe and a few other cortical areas. These may have influenced the hippocampus indirectly via the entorhinal and perirhinal cortices, but the report back is largely direct and technically non-reciprocal. The hippocampus acts as a comparator of novel and familiar stimuli (matching to expectation or prediction) and in the initiation or inhibition of behavioral strategies as appropriate to the situation. It therefore becomes linked to emotions such as anxiety and to setting appropriate motor responses (Gray, 1982).

Temporal lobe epilepsy (TLE) is a paradigm of great importance clinically. At one time this was referred to as psychomotor epilepsy, a term which emphasized the psychic and motor accompaniments of disruption during the seizure. The main pathological abnormality is hippocampal or Ammon's horn sclerosis. This may affect part or most of the hippocampus, unilaterally or bilaterally, and its pathology is gliosis associated with neuronal loss. The pathology is maximal in the CA1 and subiculum regions, the so-called Sommer's sector of the hippocampus. In spite of many years of research, hippocampal pathogenesis is still not understood, but it is known to be associated with infantile convulsions of childhood, a herald in many cases to the later development of a localization-related epilepsy (TLE). The hippocampal epilepsies comprise about 70% of TLE, other noteworthy pathologies being dysphasia and tumors, and vascular abnormalities such as angiomas. TLE is the form of epilepsy most associated with co-morbid psychopathology, and the link with the schizophrenia-like interictal psychoses has attracted most attention.

Interestingly, TLE patients that exhibit an associated schizophrenia-like psychosis are more likely to have left-sided lesions. This finding led to years of searching for hippocampal pathology and laterality effects in schizophrenia research and there is now good evidence that the hippocampus is involved in schizophrenia, either unilaterally or bilaterally, the structure being smaller than in matched controls. Some studies show lower neuronal density, others neuronal disarray; but gliosis is not a feature. Maximum changes are reported in the CA3 and 4 sub-regions (Jeste and Lohr, 1989). Thus, the intimate pathology is different in schizophrenia when compared with TLE, but the psychoses of both seem to represent the outcome of disturbed hippocampal function, beginning in the perinatal or early year or two of life.

More recent attention has been drawn to alteration of hippocampal size in the affective disorders and in some stress-related disorders such as post-traumatic stress disorder (PTSD). Adults with PTSD have been shown to have smaller hippocampal volumes bilaterally compared with trauma-exposed people without PTSD and non-traumatized controls (Kitayama et al., 2005), although whether this represents a biological predisposition from genetic and familial factors or secondary changes from the consequent metabolic changes of the stress itself is unclear. It seems established that apoptosis (cell death) occurs in the postnatal hippocampus, and that in animal models, psychotropic drugs such as antidepressants and electroconvulsive stimulation can lead to neurogenesis in the adult hippocampus. Thus, the plasticity within the structure is potentially substantial, and perhaps interlinked with its central role in the modulation of so many behaviors including memory. Steroids have been shown to lead to hippocampal neuronal damage and it has been suggested that dysfunction of neuronal plasticity or remodelling may contribute to the pathophysiology of major depression via hippocampal neuronal loss (Frodl et al., 2006).

It has been known for a long time that the hippocampus is involved in memory, and the originally described Papez circuit overlapped with a memory circuit, which included the hippocampus, fornix, and the hypothalamic mammillary bodies. Patients with TLE often report poor memory, and removal of one or the other of the hippocampi in surgery for epilepsy can lead to decrements of memory for verbal or non-verbal material depending on the side removed (verbal—left). Bilateral removal of the hippocampi leads to an amnestic syndrome, with complete failure to lay down new episodic memories (anterograde amnesia). A similar picture is seen after bilateral fornix damage, and in the Korsakoff psychosis. The latter is usually seen after an acute Wernicke's encephalopathy, following thiamine deficiency (mainly secondary to alcoholism), and damage to the mammillary bodies and medial thalamus is the outstanding neuropathology.

Hippocampal pathology is always prominent in Alzheimer's disease occurring temporally just after its earliest damage to the perirhinal and entorhinal cortices. At end stage all of these areas contain the signature pathologies of Alzheimer's disease, virtually eliminating the key circuitry that enables the hippocampus to interact with the cortex and many subcortical structures. To say that these changes literally dissect critical memory systems from the ventromedial temporal lobe is not an overstatement.

appearance of the limbic lobe, however, becomes more complex with dissection when structures like the amygdala and hippocampus are encountered. A very abrupt increase in complexity occurs with microscopic examination of limbic lobe tissue elements, which challenge even the most stalwart efforts at cytoarchitectural and phenotypic analysis. Numerous reasons account for this, but three seem paramount. First, it is obvious that the limbic lobe contains structures that are standard equipment throughout mammalian evolution but have come to be paired with newer elaborations and the "old" and "new" take slightly differing forms in different species. Second, and by definition, the edge or ring-like nature of the limbic lobe creates an interface with the isocortex that varies from one lobe to another around the ring. Limbic lobe areas that interface with the motor cortices of the frontal lobe, such as the anterior cingulate cortex, would not be expected to totally resemble a temporal limbic lobe area, such as the parahippocampal gyrus, that abuts the highly granular striate or peristriate cortex in the occipital and temporal lobes. Third, the periallocortices that form the retrosplenial, presubicular, parasubicular, and entorhinal parts of the limbic lobe are understood poorly from an ontogenetic viewpoint. Their deep layers are continuous with, and appear to be of, hippocampal allocortical origin. Connectionally, they receive from the subiculum what could be viewed as continued and strong hippocampal intrinsic connections. In this sense, it is as if the hippocampus is larger than normally appreciated, tailing off, so to speak, under the periallocortices. However, a greater problem of understanding accompanies the highly unique laminar patterns in the superficial layers that lead one to the impression that each periallocortex is a distinct multilayered cortex. The neurons comprising these layers appear to be derived from neuronal precursors that have migrated from the ventricular zone, but their final arrangement and morphology are not found elsewhere in the cortex. Hence, the neurons of the periallocortical superficial layers, critically involved in memory mechanisms, and the first affected by neurofibrillary tangles in Alzheimer's disease, are either genetically coded for uniqueness and/or represent a derived cell type from a population of early neurons largely altered before the end of corticogenesis but that survive and differentiate—that is, morph into other forms in the adult periallocortices.

All three of these features of the limbic lobe add enormously to its microscopic complexity and have thwarted interest and understanding of this cortex. For these reasons we have stripped away the confusing terminology and presented the limbic lobe in its basic forms, with more emphasis on its input from the cortex and its output back to the association cortex, the cingulate motor area, and the basal forebrain. This is largely a neural systems viewpoint unburdened to some degree from its variable cytoarchitcture.

It could be asked if Papez (1937) and MacLean (1949, 1952) didn't have it right, since both proposed systems-related concepts. The answer would have to be a lukewarm yes, acknowledging their early insight and scholarly efforts

CLINICAL BOX 13

Interictal Personality Disorder

One of the turning points in the history of neuropsychiatry was the clear delineation of the interictal personality disorder in people with temporal lobe epilepsy. Although behavioral disturbances had been described in epilepsy for centuries, these were mainly thought to be caused, for example, by head injuries the patients incurred or drugs they received. This situation persisted until Geschwind and colleagues defined some rather specific features of the interictal personality disorder (Waxman and Geschwind, 1975). These included viscosity of thinking (sluggish thought processes), hypergraphia (a tendency to extensive and elaborate writing), hyperreligiosity (excessive religious preoccupation), hyposexuality, irritability, and an increased sense of personal destiny. This profile has occasioned much controversy, yet should be considered alongside recognized personality changes that accompany other neurological disorders, such as the frontal lobe syndromes (see Clinical Boxes 5, 15 and 16).

The interictal personality disorder is found mainly in association with epilepsy arising from medial temporal lobe structures. It is seen in patients with difficult to control epilepsy and is exacerbated by additional seizures. It may emerge initially in a postictal state, possibly as a part of a postictal psychosis, in which extreme religiosity is very often a feature (Kanemoto et al., 1996). The hypergraphia is rarely of a creative nature, and some patients write poetry or religious texts. In others, what appears to be the same urge has been expressed as prolific painting.

The fact that MacLean regarded changes of behavior in temporal lobe epilepsy as an important clue to the importance of the limbic system in driving human behavior and motivation imparts historical relevance to the delineation of the interictal personality syndrome. However, the syndrome also merits considerable contemporary attention in that it reveals how a chronic yet subtle alteration of neuronal activity in a discrete brain area (in this case predominantly the hippocampus, since the pathology most often seen in such cases is hippocampal sclerosis) can lead to a skewing of the personality and the adoption of traits that influence interpersonal relationships. In many patients this has limited consequence, but, when too overt, the aversive behavioral expression can lead to a breakdown of patient care. Whereas some patients gradually become psychotic, others seek refuge in retreat. What is very relevant is that the hypergraphia and hyperreligiosity central to this syndrome reflect behaviors central to the human cultural glue. The role of emotional behavior and of limbic activity in relation to human artistic endeavors is rarely discussed, yet if art has any value it must be to arouse emotion. The interictal personality disorder is one neurological key to the discipline of neuroaesthetics (Trimble, 2007).

with existing knowledge. However, neither had appreciable insight about the input to their systems, and both failed to recognize the major output of their systems is to the basal forebrain instead of hypothalamus as they emphasized. This is the substance of our argument with them, and in our minds, the key to appreciating the role that the "new anatomy" has for understanding emotion and behavior. Hypothalamic output yields deterministic responses and behaviors be they neural, autonomic, or endocrine, and graded or not. If governed by a basic drive, such behaviors are focused and stereotyped. They can be triggered and altered by forebrain mechanisms, but overall they represent a lesser portion of our behavior. A greater portion of behavior is nondeterministic, an outcome governed by genetic predispositions, societal norms, learning, reward, and motivation. It is this aspect of behavior with which the limbic lobe and basal forebrain are largely concerned, and the anatomy described here underwrites it. It provides the cognitive base for feelings and emotions and a linkage with memory mechanisms. Thus, cortical and corticolimbic anatomy contributes to the surprisingly widespread and numerous cortical sites with functional correlates for feelings and emotions revealed in imaging studies (Damasio et al., 2000). While there are substantial individual differences and tolerances in these systems, highly negative or positive variations predispose organisms to emotional and mental abnormalities (see Clinical Box 13).

5

COOPERATION AND COMPETITION OF MACROSYSTEM OUTPUTS

5.1 BASAL FOREBRAIN FUNCTIONAL-ANATOMICAL SYSTEMS (MACROSYSTEMS) REVISITED

The ventral striatopallidum, extended amygdala, and septal-preoptic system are basal forebrain functional-anatomical systems distinguishable on the basis of cytoarchitectural and neurochemical criteria reviewed in Chapter 3. A broad variety of specific neurochemical, connectional, and functional details described by Alheid and Heimer (1988) *differentiate* basal forebrain functional-anatomical systems. Having established that macrosystems[1] are anatomically and functionally different, it was equivalently essential to point out

---

[1]Although ventral striatopallidum and extended amygdala were described in the 1970s and 1980s, the descriptor "functional-anatomical system" and synonymous term "macrosystem" emerged in subsequent papers (Heimer and Alheid, 1991; Heimer et al., 1991a).

that they also *share* an underlying organizational framework that may be referred to as a "template" (Fig. 5.1A). Drawing on the insights of numerous predecessors,[2] they thus proposed that inputs to macrosystems, largely from cortex, terminate within "input" structures comprising predominantly medium-sized, densely spiny inhibitory (i.e., GABAergic) neurons. Among such input structures would be included—for example the ventral striatum, all of extended amygdala,[3] and lateral septum.[4] These, in turn, project prominently to structures comprising mostly medium to large sparsely spined GABAergic neurons in the pallidum, extended amygdala and preoptic area that project to the cortex, via thalamo-cortical reentrant pathways, and to autonomic, somatic motor, and neuroendocrine effectors in the hypothalamus and brainstem. In addition, macrosystems establish connectional relationships with forebrain and mesencephalic cholinergic and dopaminergic neuromodulatory cell groups, providing additional thalamic and extrathalamic feedback loops to cortex and telencephalic nuclei.[5]

Basal forebrain macrosystems receive cortical and cortical-like outputs largely from structures in the greater limbic lobe, within which a high order of cortical integrative capacity is represented (see Chapter 4). One might think

[2]Alheid and Heimer (1988) cited in this regard Brodal (1947), Fallon and Loughlin (1985), Gurdjian (1928), Krettek and Price (1978), McDonald (1982), Millhouse (1986), Oertel et al. (1985), de Olmos et al. (1985), and Turner and Zimmer (1984).

[3]The extended amygdala comprises medial and central divisions. The medial division comprises medial parts of the amygdala and bed nucleus complex and associated neurons in intervening parts of the basal forebrain and stria terminalis. It is related primarily to the accessory olfactory system, cortical amygdaloid nucleus, and medial preoptic-medial hypothalamus continuum and thought to mediate neuroendocrine responses to olfactory cues, especially those concerned with reproduction and affiliation. It will not be considered further in this chapter. The central division of the extended amygdala includes the central nucleus of the amygdala, lateral parts of the bed nucleus of the stria terminalis complex, and associated neurons that extend between these structures in the sublenticular region and within the stria terminalis. It projects to the lateral hypothalamus and brainstem and, via the midline-intralaminar thalamus, especially the paraventricular nucleus, to the cortex. Purely in order to simplify the presentation, reference to extended amygdala in this chapter indicates exclusively the central division.

[4]The septal-preoptic relationships shown in Fig. 1 differ somewhat from those illustrated in Figure 15 in Alheid and Heimer's (1988) paper, where lateral septum was shown projecting to presumably GABAergic output neurons in the medial septum/diagonal band complex. The lateral preoptic area-rostral lateral hypothalamic continuum, as described in Reynolds et al. (2006), better fits the concept of a "pallidal" structure and receives dense projections from the lateral septum (Zahm, 2006). As illustrated in Fig. 1 and described in section 5.3.2, the projections of lateral septum to the medial septum/diagonal band complex, in turn, better reflect septal-preoptic relationships with the basal forebrain magnocellular (cholinergic, GABAergic, and glutamatergic) complex, of which medial septum/diagonal band represents the rostralmost extremity.

[5]The reader is referred to the earlier papers and reviews for additional detail focusing primarily on the ventral striatopallidal system and extended amygdala These include Alheid et al. (1990, 1995), de Olmos (1990, 2004), Heimer et al. (1993, 1995, 1999), de Olmos and Heimer (1999), Sakamoto et al. (1999), Heimer (2000, 2003a and b), Alheid (2003).

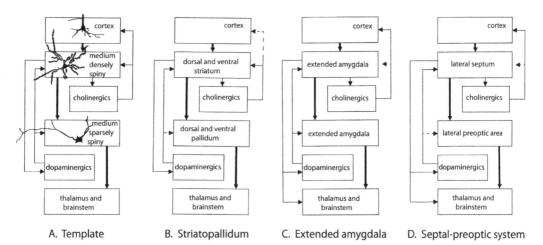

A. Template   B. Striatopallidum   C. Extended amygdala   D. Septal-preoptic system

FIGURE 5.1 Adapted from Fig. 15 in Alheid and Heimer (1988), this diagram illustrates common features shared by functional-anatomical systems. Panel A (Template) illustrates the basic organizational plan of a generic macrosystem in which cortex innervates an input structure containing medium densely spiny neurons that project to an output structure comprising medium sparsely spiny neurons that project to brainstem and thalamus, from which reentrant projections to the cortex arise. Panels B and C illustrate the three major basal forebrain macrosystems. Relationships with cholinergic and dopaminergic ascending modulatory projections are indicated. Main connections, essential to designation as a macrosystem, are indicated by thickened arrows; other arrows indicate speculative connectional relationships that in 1988 were regarded as contingent on additional study. Note that in extended amygdala input and output style neurons are designated as occupying the same structures, whereas for striatopallidum and septal-preoptic system, the input and output structures are separate. Also, note that Alheid and Heimer did not use the term *septal-preoptic system*, which has been adopted in this book (see also footnote 4).

this nothing special, knowing that the thalamus also is the recipient of very strong topographically organized outputs from the entire cerebral cortex, including the greater limbic lobe. However, the thalamus lacks outputs that descend toward motor effectors and, moreover, the corticothalamic projection system is directly reciprocated by equivalently robust thalamocortical connections. Thus, because the corticothalamic and thalamocortical projections both are excitatory, activity in interconnected parts of thalamus and cortex must, to some significant extent, be interlocked. In contrast, cortical outputs to "striatal" neurons in the macrosystems are "one way"—that is, the striatum lacks direct projections back to cortex. Nor do striatal neurons project to the thalamus, having instead only intrinsic and descending projections. Consequently, cortically acquired information conducted to macrosystem input structures may be subjected there to "biasing" imposed by the macrosystem intrinsic circuitries and other macrosystem afferents. Because macrosystems do not project reciprocally to the cortex, they are favorably situated so as to engage mechanisms to modify cortically acquired data and disseminate it in altered form to a variety of designated brain structures concerned with the synthesis of behavior.

Again, one may be inclined to suggest that there is nothing special about all of this. It has been known for decades that the input-output organizations of the basal ganglia and cerebellum are precisely as is described in the preceding paragraph. But this is precisely the point: Basal forebrain macrosystems subserve neural mechanisms not fundamentally different than those subserved by the basal ganglia and cerebellum. Thus, it is pointless to put them in a limbic system, unless one would do the same with the basal ganglia and the cerebellum, but then the system is no longer "limbic" and truly begins to approximate the entire brain, as Brodal predicted it would.

Nevertheless, some properties of basal forebrain macrosystems distinguish them from basal ganglia and cerebellum. As noted, basal forebrain macrosystems are output targets of the greater limbic lobe, which happens to be the part of the cortical mantle most richly interconnected with extrathalamic subcortical structures, such as the lateral hypothalamus and a variety of transmitter-specified (i.e., cholinergic, dopaminergic, serotoninergic, histaminergic, etc.) ascending neuromodulatory projections. The greater limbic lobe houses all of the cortical structures most closely associated with emotional expression (laterobasal-cortical amygdala and medial prefrontal/anterior cingulate areas) and memory and learning processes (entorhinal area and hippocampus). Interestingly, all of these limbic lobe structures project more or less robustly to all of the basal forebrain macrosystems. Lateral septum, for example, gets inputs predominantly from hippocampus supplemented by fewer from the prefrontal cortex and basolateral amygdala. The extended amygdala receives its most prominent input from the basolateral amygdala and a lesser complement from the prefrontal cortex and hippocampus. The accumbens receives more or less equivalently robust inputs from the prefrontal cortex, basolateral amygdala, and hippocampus. This arrangement suggests that each of the several macrosystems gets information derived from much of the greater limbic lobe—and such information is likely to be reprocessed by each macrosystem to a different effect and, thus, a different impact on behavioral synthesis.

Thus, one readily sees that we intend in this chapter to develop a model in which neural output derivative of high order, cortically derived cognitive representations is channeled from the greater limbic lobe through several macrosystems, each of which may read messages differently and redistribute them in altered form to the brainstem (Fig. 5.2A) and, via reentrant circuits, back to the cognitive apparatus (Fig. 5.2B). We will move in this direction, first by discussing the organization and distribution of macrosystem outputs. Then, because we think it is fundamental to the model that at some level some degree of mutual segregation of the neural processing functions of the macrosystems must exist, we will discuss anatomical organization consistent with the

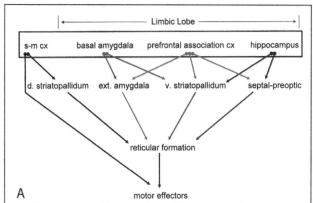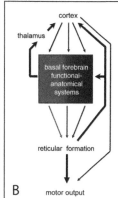

FIGURE 5.2 Connectional relationships of basal forebrain macrosystems. A. Outputs from greater limbic lobe structures, including the basal amygdala, prefrontal cortex (cx), and hippocampus diverge to innervate multiple macrosystems, including extended (ext.) amygdala, ventral (v.) striatopallidum, and septal-preoptic system, which, together with the dorsal striatopallidum, give rise to outputs that converge in the reticular formation. Outputs from the reticular formation and cortex, in turn, converge upon motor effectors. B. Replicates features illustrated in A, but, in addition, emphasizes reentrant pathways returning (1) to the cortex via the thalamus and (2) to the cortex and/or deep telencephalic nuclei (the macrosystems themselves) via ascending modulatory projection systems. Additional abbreviation: s-m—sensorimotor. See color plate.

proposed segregation. Ultimately, we will propose that the functional-anatomical properties of macrosystems favorably position their outputs to compete and cooperate to influence cognitive, emotional, and motor components of behavioral synthesis. However, in order to consider how macrosystem outputs might influence behavioral synthesis, it is first necessary to discuss some functional-anatomical characteristics of the reticular formation of the brain.

5.2 RETICULAR FORMATION AND BEHAVIORAL SYNTHESIS

The lower brainstem maintains essential vegetative functions, such as respiration, heartbeat, and blood pressure, and, via ascending projections, exerts essential controls on cortical and thalamocortical activation—that is, on arousal and, hence, sleeping, waking, discerning, and attending (Blessing, 1997). In addition to this, the brainstem provides essential descending projections to somatic motor pattern generators in the spinal cord. A rich legacy of experimentation carried out during the previous two centuries by numerous investigators revealed that many diverse animal forms from amphibian through higher mammal retain substantial capacity for integrated, coordinated motor control following complete transections of the neuraxis just rostral to the mesencephalon.[6] The mesencephalic rat, in particular, is capable of quite complicated, seemingly goal-directed patterns of movement (Woods, 1963). Thus, an organism's basic repertoire of autonomic, postural, and locomotor actions are executed effectively by brainstem-spinal interactions in the absence of the cerebral hemispheres and diencephalon. The neural substrate regarded as most essential in supporting this broad array of complex functions is the brainstem reticular formation.[7] Motivated adaptive behavior, however, requires that these essentially autonomous actions of the brainstem be modulated by descending influences (Harris, 1958).

The functional capacities of the brainstem reticular formation may even extend well beyond what was described in the preceding paragraph. Take, for example, the pedunculopontine tegmental nucleus (PPTg), a constituent of the reticular formation located in the mesopontine tegmentum and endowed with particularly rich ascending and descending connections (Mesulam, 1995). Winn (1998, 2006) has written persuasively to the effect that bilateral lesions of the PPTg produce a behavioral syndrome in rats involving disinhibition, impulsiveness, and perseveration that much resembles behavior observed following bilateral lesions of the medial prefrontal cortex, one of the main constituents of the greater limbic lobe (see Clinical Box 14). Thus, any explanation of neural mechanisms underlying motivated behavior must address not only the nature of descending controls but also the brainstem circuitry in which such controls are exerted and the reentrant pathways to the cognitive apparatus that originate there.

[6]There exists an extensive literature going back to the latter part of the 19th century detailing studies of behavior in a variety of vertebrates following complete surgical transections at various levels of the neuroaxis. In general, postural and locomotor functions were rather well retained in the *cerveau isole* preparation, i.e., transections in front of the rostral mesencephalon. Following such lesions the teleost fish "swims constantly in a straight line, but avoids obstacles. . . . fails to school, fight or mate" (Ferrier, 1876), the reptile "remains in position in which it is placed, but can walk or run in a perfectly coordinated manner when stimulated" (Goldby, 1937), the guinea pig "responds to nuzzling by male by assuming a receptive posture" (Dempsey and Rioch, 1939) and the cat "assumes normal species typical posture for walking, running, urination and defecation, does not clean itself" (Bard and Rioch, 1937). In the dog, "full bladder, hunger, stimulation result in rising and running around with depressed head, running into all obstacles. . . . lacks spontaneity. . . . lived three months, learned nothing" (Dresel, 1924). The rat, however, "nibbled at objects, edible and inedible . . . grasped pipettes with their forepaws . . . took care of their coats . . . and . . . displayed typical rodent defense behavior (vocalization, attempts to escape, use of claws and teeth) with accurate localization of the stimulus. . . ." (Woods, 1964).

[7]Included among many authors that have written on the brainstem reticular formation are Meesen and Olszewski (1949), Moruzzi and Magoun (1949), Brodal (1957), Olszewski (1957), Olszewski and Baxter (1954), Scheibel and Scheibel (1958), Leontovich and Zhukova (1963), Ramon-Moliner and Nauta (1966), Nauta and Feirtag (1979), Jones (1995), and Blessing (1997).

> ## CLINICAL BOX 14
>
> ### Pontine and Mesencephalic Lesions
>
> In classical neurology, the possibility was not often considered that pathology from pontine lesions may lead to clinical neuropsychiatric syndromes. However, several different syndromes are recognized following lesions to neuronal structures in the hindbrain, often combined with evident loss of function of one or more of the cranial nerves. Lesions apparently involving the reticular formation can markedly alter behavior, including disruptions of the sleep-wake cycle or even such dramatic states as a locked-in syndrome. Peduncular hallucinosis, characterized by visual hallucinations, often with some clouding of consciousness, has been reported following a variety of brainstem lesions involving substantia nigra reticulata, red nucleus, and the tegmentum of the pons and lower midbrain. Behavioral changes secondary to central pontine myelinolysis have been described, possibly related to involvement of ascending projections from the pontomesencephalic reticular formation (Varma and Trimble, 1997). Further, two patients have been described, who, following brainstem injuries, developed behavioral changes resembling a frontal syndrome. Both patients had neuro-ophthalmological signs indicative of lesions of the rostral brainstem, which were confirmed by CT scanning, and it was suggested that the behavioral changes were related to the disruption of nearby limbic, hypothalamic, and reticular structures (Cummings and Trimble 1981).

The brainstem reticular formation, through its entire length, is pervaded by a core of varyingly dispersed neurons of various sizes with long, aspiny, sparsely branched dendrites. Broad dendritic domains extensively overlap each other in the resulting reticulum-like milieu, often in orientations tending to be perpendicular to abundant highly collateralized extrinsic and intrinsic fiber systems that ascend and descend among them and provide them with varyingly sparse or dense inputs. Nuclei such as, for example, cranial nerve motor and sensory nuclei are embedded within the reticular matrix formed by the dendrites and axons of these so-called "isodendritic" neurons (Ramon-Moliner and Nauta, 1966). Because the neurons comprising the nuclei frequently extend long sparsely branched dendrites far into the reticular matrix, however, it frequently is not a trivial task, if possible at all, to objectify what is nucleus and what is isodendritic infrastructure. The origins of the efferents and terminations of the afferents of such brainstem nuclei also tend to be vaguely delimited. Thus, convincing models of structure-function relationships in the brainstem reticular formation have yet to emerge.

Lacking a comprehensible conceptual framework for structure-function relationships in the reticular formation, however, did not deter neuroscientists from utilizing Ramon-Moliner and Nauta's (1966) "isodendritic core of the brainstem" as a basis for a more liberal estimation of its extent.[8] Indeed, the entire rostrocaudal extent of the lateral hypothalamus and preoptic region and the basal forebrain magnocellular formation comprise neurons that exhibit dendritic morphology and connectional relationships indistinguishable from those observed in the isodendritic core. Furthermore, brainstem aminergic and cholinergic projection neurons that give rise to the so-called long ascending modulatory projections also have long, simple dendrites that are innervated from multiple sources and axons that ascend (and perhaps also descend) over substantial distances to ramify among other structures in the reticular formation, rarely emitting collaterals. Only their abundant projections into the telencephalon might be regarded as distinguishing. Even if these cell groups were to be conservatively excluded from reticular formation, their neuroanatomical organization and relationships predict that they process information in a manner resembling what occurs in the isodendritic core of the brain.

5.3 MACROSYSTEM OUTPUTS

Outputs from the macrosystems form two contingents: reentrant and descending (Fig. 5.2B). Some of the outputs initially descend, only to contact neurons in the thalamus possessing axons that loop back into the cortex and form part of the so-called reentrant pathways. Other macrosystem outputs integrate into the circuitry of the brainstem reticular formation, descending in the medial forebrain bundle to pass in relation to a variety of structures of which all have been regarded by various students of the brainstem either as formal parts of the reticular formation or intimately affiliated with it. Some of these, named in descending order, include the subcommissural (beneath the anterior commissure) and preoptic regions, lateral and medial hypothalamus, sublenticular (beneath the globus pallidus) region, subthalamic region, epithalamus (lateral habenula), ventral mesencephalon (ventral tegmental area, substantia nigra, retrorubral field), mesopontine tegmentum (pedunculopontine and laterodorsal tegmental nuclei, parabrachial region), and more caudal brainstem sites (e.g., nucleus of the solitary tract, dorsal vagal complex). In the case of some macrosystem outputs, such as, for example, those from the accumbens core, the fibers course among these structures in tightly fasciculated fashion, giving off robust "bursts" of terminations only within or in the vicinity of particular structures. In other cases, good examples being the accumbens shell and

---

[8]Included among authors that have advocated an expanded conceptualization of the reticular formation are Scheibel and Scheibel (1958), Leontovich and Zhukova (1963), Ramon-Moliner and Nauta (1966), Shute and Lewis (1967), Ramon-Moliner (1975), McMullen and Almli (1981), Arendt et al. (1995), Jones (1995), Mesulam (1995), and Gritti et al. (1997).

extended amygdaloid outputs, the fibers are more loosely, if at all, fasciculated and have countless varicosities and short collaterals presumed to be sites of synaptic or parasynaptic transmission that involve not only defined structures but also points of indeterminate neural organization all along the course of the medial forebrain bundle. Descending macrosystem outputs pass in relation to a variety of the aforementioned transmitter-specified ascending neuromodulatory projection systems that comprise the other part of the reentrant pathways. These consist of variously loosely organized aggregations of neurons with long axons that ascend to various diencephalic and telencephalic sites, including the thalamus, deep telencephalic nuclei (i.e., the macrosystems themselves), and cortex, where they arborize extensively, giving rise to varyingly dense fields of terminations. Some of the transmitter/modulators represented by such ascending neuromodulatory projections include acetylcholine, GABA, glutamate, orexin, histamine, dopamine, norepinephrine, serotonin, and epinephrine. In this account, we will provide some details only about the cholinergic and dopaminergic reentrant pathways, but the reader should be aware that outputs descending from the macrosystems surely interact with all.

As a consequence of their descent into this poorly defined organization, it is exceedingly difficult to conceive of precise, definitive mechanisms that outputs from the forebrain, including from the macrosystems, might engage to orchestrate purposeful, adaptive behavior from the extensive and sophisticated repertoire of autonomous, albeit restrictively programmed, hindbrain-spinal motor routines. But it is equivalently hard to dispel the notion that this is precisely what happens, and it seems advisable to presume that we are at present unable to perceive the relevant functional-anatomical relationships rather than to think they don't exist. However, it is not accurate to say that we are entirely ignorant of organizational features that characterize descending macrosystem outputs. To the contrary, some basic patterns of organization have been recognized within the connectivity relating the macrosystems to their output targets in the reticular formation and thalamus, and it seems reasonable to expect that further concerted efforts will lead incrementally to the emergence of a more comprehensive conceptual framework. In forthcoming sections we will comment on some of what has been shown so far.

5.3.1 Rostrocaudal Biasing of Macrosystem Outputs

As has been mentioned, the outputs from the basal forebrain macrosystems either project directly to the thalamus, which gives rise to thalamocortical projections, or descend into the basal forebrain and brainstem for varying distances, where they engage the circuitry of the reticular formation, including, via direct projections or interneuronal relays, groups of neurons that give rise to ascending neuromodulatory projections. The transthalamic pathways and ascending modulatory systems contribute to the regulation of arousal and attentional processes and influence cognition, thus modulating cortical components of behavioral synthesis. Macrosystem outputs descending within the medial forebrain bundle into the basal forebrain and diencephalic and brainstem reticular formation influence the synthesis of purposeful, coordinated motor activity and behavior-appropriate autonomic and neuroendocrine responses. Ouputs from all of the macrosystems converge extensively within these descending trajectories but also exhibit significant levels of projection specificity.

Thalamocortical reentrant pathways from the dorsal striatopallidum,[9] via the thalamic ventral tier nuclei, on one hand, and the midline-intralaminar complex, on the other, are prominent by comparison to the more moderate downstream projections to the vicinity of the pedunculopontine tegmental nucleus, lateral habenula, and deep layers of the superior colliculus, of which all are regarded as part of or affiliated with the reticular formation and likely to impact brainstem motor function (Fig. 5.3A). Ventral striatopallidum also is characterized by dual transthalamic reentrant pathways that traverse the mediodorsal nucleus and midline-intralaminar nuclei, but the descending pathway from ventral striatopallidum is more substantial, owing to its robust projections to the lateral hypothalamus, which, in turn, has strong projections to reticular formation structures in the mesopontine tegmentum and caudal brainstem (Fig. 5.3B). Extended amygdala and, presumably, septal-preoptic system are characterized by even more caudally biased output systems, with uniformly direct and robust downstream projections to lateral hypothalamus and the caudal brainstem (Fig. 5.3C). Lacking rostropetal projections via thalamic relay nuclei, extended amygdala and septal-preoptic system have only modest projections to the thalamic midline and intralaminar groups.

These patterns of connectivity fit generally with functions attributed to the macrosystems. Dorsal striatopallidum, for example, is implicated in voluntary, finely coordinated appendicular motor function and the cognitive processing that underlies it, consistent with the strong transthalamic dorsal striatopallidal reentrant connectivity. Ventral striatopallidum is thought to contribute to motivational processes and locomotor activation consistent with approach behavior, which in part reflect cognitive functioning but also involve a more automatic component. The lesser ventral striatopallidal transthalamic connectivity and greater downstream projections, particularly to the lateral hypothalamus and ventral tegmental area, are appropriate to these considerations. In contrast, extended amygdala and likely the septal-preoptic system, are involved in the adjustment of motor set points in accord with emotional state,

---

[9]We will routinely consider dorsal striatopallidum in our discussions of the outputs before turning to the basal forebrain macrosystems, insofar as the "generalized template" to which all of the macrosystems conform is also represented by dorsal striatopallidum, which makes it a consistently instructive starting point.

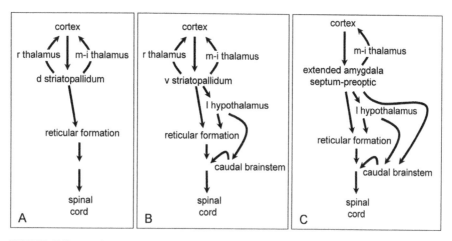

FIGURE 5.3 Further connectional relationships of dorsal (d) and ventral (v) striatopallidum, extended amygdala, and septal-preoptic system emphasizing that functional-anatomical macrosystems extend from cortex to the brainstem reticular formation to establish continuity with motor networks in the brainstem and spinal cord. Relay (r.) and midline-intralaminar (m-i) reentrant paths to the cortex via thalamus (th.) are indicated. Dorsal striatopallidum has relatively strong relationships with the cortex as compared to its downstream projections, whereas extended amygdala and septal-preoptic system lack a reentrant pathway utilizing a thalamic relay nucleus and are more strongly related to the caudal brainstem. The strength of connectional relationships of ventral striatopallidum with cortex and caudal brainstem is intermediate between those of dorsal striatopallidum and extended amygdala/septum-preoptic.

as in, for example, fear-associated freezing and conditioned startle, consistent with the prominent descending projections of these macrosystems to caudal brainstem somatomotor and autonomic effector sites. As will be described in a forthcoming section, a case can be made that all of the macrosystems also contribute to the content of emotional states through their influences on the various transmitter-specific ascending modulatory projections.

5.3.2 Outputs to the Ascending Cholinergic System

Dorsal Striatopallidum

A possibility for projections from the dorsal striatopallidum to directly contact basal forebrain corticopetal cholinergic neurons would seem limited to those embedded within the substance of the globus pallidus and internal capsule (Henderson, 1997; Smiley and Mesulam, 1999), which project broadly to parts of the frontal lobe that give rise to the cortico-striato-pallidal innervation of the district in which they lie, thus completing a loosely "closed" cortico-striato-cholinergic neuron-cortical circuit (Fig. 5.4A). The pathway

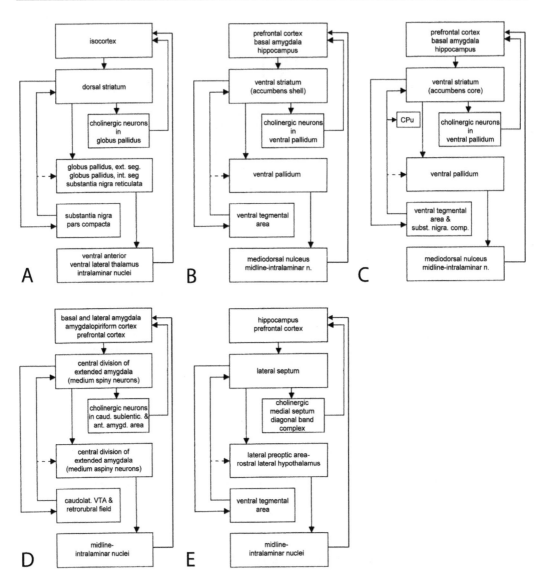

FIGURE 5.4 More specific depiction of macrosystem relationships shown in the format of Fig. 1, an adaptation of the original diagram of Alheid and Heimer (1988). See sections 5.3.2 and 5.3.3 for details.

descending from dorsal striatum otherwise encounters no significant aggregations of cholinergic neurons. However, the dorsal striatopallidal output nuclei, including the substantia nigra reticulata and medial segment of the globus pallidus, project to the vicinity of the pedunculopontine tegmental nucleus (PPTg), which contains a dense aggregation of rostrally projecting cholinergic neurons. This apparently is also the case for the subthalamic nucleus, an important structure in the intrinsic circuitry of striatopallidum. It must be pointed out, however, that these projections actually deploy most of their terminations in the mesopontine region just medial to main body, or so-called pars compacta, of the PPTg (Rye et al., 1987), which gives rise to most of its ascending cholinergic projections. Thus, if significant interactions exist between the descending striatopallidal projections and cholinergic PPTg projection neurons, they are likely to be modest and/or polysynaptic in character. As we will describe in forthcoming paragraphs, this point applies to projections descending to the PPTg from all of the telencephalic macrosystems, of which all converge in this part of the mesopontine tegmentum.

Ventral Striatopallidum

The ventral pallidum is occupied by cholinergic neurons (Fig. 5.5), which are contacted by the axons of ventral striatal neurons (Grove et al., 1986; Záborszky and Cullinan, 1992) and project to the laterobasal amygdaloid complex (Carlsen et al., 1985), which in turn projects strongly to ventral striatum (Figs. 5.4B and C). Thus, these cholinergic neurons also are a link in a "closed" neuronal circuit. Beyond the ventral pallidum, however, projections from the ventral striatum and ventral pallidum encounter no further basal forebrain cholinergic projection neurons. Nor does ventral striatopallidum appear to have a direct projection to the PPTg. That is, excepting a few stray fibers, the descending pathway from ventral striatopallidum ends in the ventral mesencephalic tegmentum and periaqueductal gray well forward of the PPTg (Zahm et al., 1999; 2001; Gastard et al., 2002), ruling out significant monosynaptic influences. However, ventral striatum and ventral pallidum both have robust terminations in the lateral hypothalamus, which in turn projects robustly to the mesopontine tegmental corridor medial to the PPTg (Nauta and Domesick, 1978), wherein outputs from dorsal striatopallidum were noted in the preceding section.

Extended Amygdala

The extended amgydala projects strongly into a part of the basal forebrain cholinergic complex located in the caudalmost part of the sublenticular region and adjacent anterior amygdaloid area (Fig. 5.5). In the rat, this is the only substantial population of basal forebrain cholinergic neurons distributed

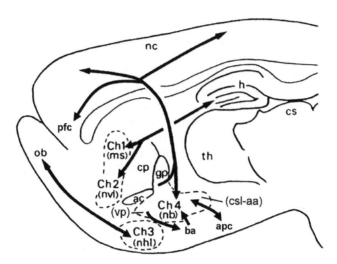

FIGURE 5.5 Loosely reciprocal connections of the cortex and magnocellular basal forebrain (cholinergic, GABAergic, glutamatergic) complex adapted from Fig. 6 of Mesulam (1983). The Ch4 cholinergic group of Mesulam (1983), also known as the nucleus basalis (nb), is depicted as extending to include parts of the globus pallidus (gp), ventral pallidum (vp), and caudal sublenticular-anterior amygdaloid region (csl-aa, see also Basic Science Box 1). It is not certain that cholinergic neurons in the GP and VP receive direct projections from the cortex, which may be true for other parts of the complex. Additional abbreviations: apc—amygdalopiriform transition cortex; ac—anterior commissure; ba—basal amygdala; Ch1, Ch2, Ch3—cholinergic groups of Mesulam (1983); cp—caudate-putamen; cs—superior colliculus; h—hippocampus; ms—medial septum; nc—neocortex; nhl—nucleus of the horizontal limb of the diagonal band; nvl—nucleus of the vertical limb of the diagonal band; ob—olfactory bulb; pfc—prefrontal cortex; th—thalamus. Modified from Fig. 6 in Mesulam et al., 1983, with permission.

coextensively with the extended amygdala itself or outputs arising from it (Gastard et al., 2002). This group of cholinergic neurons projects to an area of the temporal lobe cortex known as the amygdalopiriform transition area (Zahm et al., 2004), which in turn strongly innervates the entire central division of the extended amygdala (Shammah-Lagnado and Santiago, 1999). Consequently, the afferent and efferent connections of this distinctive part of the magnocellular corticopetal group, together with the amygdalopiriform transition cortex, comprise a "closed" circuit (Fig. 5.4D). In some respects, the cholinergics innervated by the extended amygdala are special and, thus, provide insights relevant to the overall organization of macrosystem-basal forebrain cholinergic interactions (see Basic Science Box 6). Outputs from extended amygdala to the vicinity of cholinergic cell groups in the PPTg converge with those from dorsal and ventral striatopallidum in the territory lying medial to the PPTg, where relatively few cholinergic neurons are present (Gastard et al., 2002).

BASIC SCIENCE BOX 6

An Instructive Corner of the Rat Magnocellular Basal Forebrain

It is common for the corticopetal (projecting to the cortex) cholinergic neurons of the magnocellular basal forebrain to be regarded as comprising a relatively homogeneous phenotype, albeit one represented by neurons scattered through the basal forebrain in groups, clusters, and individually, as was described in Chapter 3 (see also Záborszky et al., 1991, 1999). However, mounting evidence indicates that basal forebrain cholinergic neurons instead comprise a heterogeneous collection of different kinds of neurons, the cholinergic phenoptype being just one common thread. For example, Heckers et al. (1994) reported that cholinergic neurons projecting to the amygdala lack the p75 low affinity nerve growth factor receptor, which, previously, all basal forebrain cholinergic neurons were thought to express.

Another subgroup of cholinergic neurons, which happens to comprise those contacted by extended amygdala outputs, also is morphologically and neurochemically different from other basal forebrain cholinergic neurons. This subpopulation, described in rat, occupies the so-called caudal sublenticular region and anterior amygdaloid area (csl-aa in Fig. 5.5), in the caudalmost part of the Ch4 district of Mesulam et al. (1983). Cholinergic neurons comprising this group are significantly smaller than those in other parts of the magnocellular basal forebrain complex (Gastard et al., 2002) and exhibit significantly weaker expression of cholinergic markers, such as choline acetyltransferase and vesicular acetylcholine transporter (Loopuijt et al., 2003; Zahm et al., 2004). Like the amygdala-projecting neurons described by Heckers et al. (1994), those in the caudal sublenticular/anterior amygdaloid area do not express the p75 low affinity nerve growth factor receptor (Loopuijt et al., 2003). Furthermore, they exhibit distinct ultrastructural features and synaptology, by comparison to cholinergic neurons outside of the extended amygdala (Loopuijt and Zahm, 2006). For example, prominent stacks of rough endoplasmic reticulum (rer) that occupy the cytoplasm of large basal forebrain cholinergic neurons are absent in the caudal group of small cholinergics, which exhibit instead poorly organized, inconspicuous rer. Furthermore, cholinergic neurons within the caudal extended amygdala have significant numbers of synapses, largely with symmetrical morphology and most likely reflecting inputs from the GABAergic intrinsic network of the caudal part of extended amygdala, in which they lie embedded. This distinguishes them from other cholinergics, which typically have few synaptic contacts on the perikarya (cell bodies) and proximal dendrites (Záborszky and Cullinan, 1992; Henderson, 1997; Jolkkonnen et al., 2002). Such a conspicuous difference in the pattern of inputs obviously has implications for the regulation of neuronal activity by afferents.

Outputs from the CeA in the monkey apparently arborize more widely among basal forebrain cholinergics, reportedly extending throughout the nucleus basalis and horizontal limb of the diagonal band (Price and Amaral, 1981). However, it may be that the monkey basal forebrain exhibits a more widely distributed population of neurochemically and structurally distinct cholinergic neurons homologous to those that in the rat are sequestered within the caudal part of Ch4. Were a distinct phenotype of extended amygdala-associated cholinergics simply found to be more broadly disseminated within the monkey basal forebrain, it is likely that the innervation (from extended amgydala) would be found to follow them. This seems to reflect another situation wherein the rat brain may serve a heuristic role, as it so frequently has, in encouraging further investigation of a specific issue in the primate, in this case germaine to the basal forebrain cholinergics.

Septal-Preoptic System

Connectivity involving a robust projection from lateral septum to the medial septum-diagonal band complex, which contains a dense aggregation of cholinergic magnocellular neurons that project to the hippocampus, has been the subject of much interest (e.g., Swanson et al., 1987; Leranth and Frotscher, 1989; Jakab and Leranth, 1995). Insofar as the hippocampus also provides the major cortical input to the lateral septum, this is yet another example of a circuit involving a macrosystem input structure (lateral septum) projecting selectively to a restricted segment of the cholinergic magnocellar basal forebrain (medial septum) and receiving input from a cortex (hippopcampus) innervated by those cholinergic neurons (Fig. 5.4E). This kind of circuitous pattern of organization is quite consistent across the several basal forebrain macrosystems that have been described (Fig. 5.4A–E).

Macrosystem Outputs to Basal Forebrain Cholinergics Summarized

The main message emerging from a consideration of macrosystem projections to the basal forebrain cholinergic cell group revives and extends a concept articulated by Laszlo Záborszky (1991) with a focus on striatopallidum. The idea (extended to all of the macrosystems in this account) is that a macrosystem innervates a particular segment of the basal forebrain cholinergic neurons that in turn projects to the cortex (or cortical-like structure, in the case of the laterobasal amygdala) that innervates the macrosystem, thus forming a more or less "closed" circuit (Fig. 5.4). It is as if the particular complement of cholinergic neurons associated with each macrosystem is integral to the particular functional character of that macrosystem and might thus be regarded as part of its intrinsic circuitry. However, a few additional points need to be added in here in order to fine-tune the portrayal of these circuits to better approximate reality. First, it should always be kept in mind that, when we reference that one system "projects to" another, as in, for example, a cortex projecting to a segment of the cholinergic cell group, this does not necessarily mean that a monosynaptic connection exists. To demonstrate or rule out such a thing conclusively requires heroic efforts with the electron microscope that generally have yet to be exerted. Consequently, unless specifically stated to the contrary, "projecting to" generally should be taken to mean "projecting to the vicinity of" in the sense that the precise nature of the synaptic interactions remain unknown. These considerations are particularly germaine in the case of the basal forebrain and mesopontine tegmental cholinergic neurons, which are possessed of very long radiating dendrites that may be differentially innervated depending on whether one looks proximally (nearer the cell body) or distally (further from the cell body).

A second caveat accompanies our repeated reference to the basal forebrain cholinergic neurons as contributing to "closed" circuits. Since the spread of projections, particularly those of the ascending modulatory projections looping back to the cortex, typically is quite broad and thus likely to involve widespread cortical areas outside of the circuit origin, all such circuits described in the preceding paragraphs likely include fibers that do not end up in the cortex of origin and, thus, reflect an "open" pathway rather than a "closed" circuit (Fig. 5.4B–E). Furthermore, all of the so-called "closed" circuits are joined by corticostriatal inputs originating in cortical or cortical-like (cortical-laterobasal amygdaloid) areas distinct from the "main" input to a particular macrosystem. For example, while the main cortical input to the lateral septal-medial septal-hippocampal loop is the hippocampus, lateral septum also recieves moderate inputs from prefrontal cortex and, to a lesser extent, laterobasal amygdala. The impact of these "extrinsic" inputs to the circuit is nonetheless conveyed to hippocampus, the main site of termination of the medial septal cholinergic neurons. The accumbens, in contrast, receives strong projections from hippocampus, laterobasal amygdala, and prefrontal cortex, but since the cholinergic neurons embedded in the ventral pallidum project largely to the laterobasal amygdaloid complex, it is likely that the impact of the hippocampal and prefrontocortical inputs to this circuit are exerted, via the cholinergic neurons in ventral pallidum, on the laterobasal amygdaloid complex. Thus, the so-called "closed" circuits are at best only partially closed and contribute generously to integrative mechanisms. Also, as was noted in Chapter 3, a number of robust circuits have been identified that involve direct projections from the cerebral cortex to subpopulations of basal forebrain cholinergic neurons, which, in turn, project in significant part back to the cortex of origin (Mesulam, 1983a, b; Saper, 1984; Záborszky et al., 1991, 1999). The extent to which these circuits merge and interdigitate with those involving macrosystems and basal forebrain cholinergic neurons is not well established (Fig. 5.5). Thus, we conclude by reiterating that each macrosystem appears to be preferentially related to a macrosystem-basal forebrain cholinergic reentrant circuit that has a particular "character" imparted by a variety of factors, such as the topography and somatodendritic architecture of the involved cholinergic neurons and the extrinsic inputs to the circuit. The different character of cholinergic reentrant circuits might be expected to contribute significantly to the distinct functional characters of the macrosystems.

Macrosystem Outputs to Mesopontine Cholinergics Summarized

Whereas macrosystem-basal forebrain cholinergic reentrant loops are organized in a parallel, if not precisely segregated, fashion reminiscent of the cortico-basal ganglia-thalamocortical loops (see Chapter 3 and following), the macrosystem outputs that descend to the vicinity of the mesopontine tegmental cholinergics in the PPTg merge and interdigitate extensively there. From the vicinity of the PPTg and LDTg, where such convergence is particularly striking, not only cholinergic but also GABAergic and glutamatergic projections ascend toward the ventral mesencephalon, basal forebrain, and thalamus, where they terminate in particularly robust fashion. Furthermore, robust caudalward projections from the vicinity of the PPTg to brainstem motor effector sites also have been reported. Thus, the neuroanatomy suggests that convergent macrosystem outputs gain access in the mesopontine tegmentum, via either direct connections or interneuronal relays, to robust projections that ascend to the forebrain and descend to brainstem somatomotor and autonomic effector sites.[10]

5.3.3 Outputs to the Mesotelencephalic Dopaminergic System

Like the cholinergic magnocellular basal forebrain system, the ascending dopaminergic projection system arises from a collection of transmitter-specified neurons dispersed across a broad expanse of brain—in this case, the ventral mesencephalon. Three major subdivisions of the mesotelencephalic dopaminergic neuronal pool are recognized—the more laterally positioned substantia nigra pars compacta (also called the A9 group), a medial ventral tegmental area (A10), and a caudal retrorubral field (A8). These subdivisions merge with each other across graded transitions, and each has further subdivisions that do the same. The relationships of the basal forebrain macrosystems to the

---

[10]Included among many authors who have written on the connectional relationships of the PPTg and LDTg are Hallanger and Wainer (1988), Inglis and Winn (1995), Oakman et al. (1999), Satoh and Fibiger (1986), and Woolf and Butcher (1986). Those who have addressed more specifically the descending projections to the pontine and medullary reticular formation and spinal cord include Inglis and Winn (1995), Mitani et al. (1988), Rye et al. (1988), and Semba (1993).

ventral mesencephalic dopaminergic neurons is reminiscent of the projections of the macrosystems to the basal forebrain cholinergic neurons, in the sense that none innervates the entire complex of dopaminergic neurons.

Dorsal Striatopallidum

Dorsal striatopallidum projects strongly to the dopaminergic neurons of the substantia nigra pars compacta and massively to the subjacent substantia nigra pars reticulata, which contains clusters of dopaminergic neurons and the ventrally extending dendrites of dopaminergic neurons in the pars compacta (Fig. 5.4A). Many details are known about the topography of this projection, and, particularly, the specific connections of neurochemically distinct dorsal striatal compartments—the so-called patches (striosomes) and matrix. The globus pallidus and entopeduncular nucleus (medial pallidal segment) project largely to the substantia nigra reticulata, but also contact dopaminergic neurons and dendrites therein. These relationships will not be described in greater detail here,[11] except to point out that sites in the dorsal striatum and substantia nigra pars compacta are more or less reciprocally interconnected—that is, "closed" circuits are formed by the striato-nigro-striatal and striatopallido-nigrostriatal connectivities. In this regard, however, it is important to emphasize again (as was seen for the cholinergic reentrant pathways to cortex) that the topographies of the dopaminergic reentrant projections to the dorsal striatum are sufficiently broad that substantial parts of the return end in districts adjacent to those in which the striatonigral link of the circuit arises and thus reflect "open" pathways rather than "closed" circuits (see following and Haber et al., 2000).

Ventral Striatopallidum

The accumbens and related parts of ventral striatum and ventral pallidum project strongly to the ventral tegmental area (VTA). These projections exhibit a broadly overlapping mediolateral topography and abundant synaptic contacts with dopaminergic neurons. Outputs from shell of the accumbens are densest in the medial three-quarters of the VTA, which in turn project back to the shell, again forming a "closed" circuit, whereas those from the accumbens core occupy the lateral three-quarters of the VTA, which projects back to core (Fig. 5.4B). Core projections and also extend lateralward through much of the substantia nigra pars compacta, such that activity in the accumbens core

---

[11]These relationships are described in a large review of the basal ganglia by Heimer et al. (1995).

should be expected to have implications for dopamine release in the dorsal striatum.[12]

Extended Amygdala

Extended amygdala outputs pass through the VTA and medial SNc with little evidence of synaptic differentiation and then, somewhat like those from the accumbens core, turn lateralward enroute to the lateral part of the SNc and suprajacent retrorubral field, where the projection exhibits a robust "burst" of varicosities (Fig. 5.6). This part of the retrorubral field, together with some projections from the ventrolateral periaqueductal gray, provides much of the dopaminergic innervation of the central division of the extended amygdala (Hasue and Shammah-Lagnado, 2001). Thus, we point out another "closed" circuit inter-relating a part of the ventral mesencephalic dopaminergic projection system and the macrosystem that innervates it (Fig. 5.4D).

Outputs to the Mesotelencephalic Dopaminergics Summarized

We can conclude this section with an acknowledgement that the lateral septum projects moderately to and receives dopaminergic projections from the rostromedial part of the VTA (Fig. 5.4E), thus rounding out the general observation that all of the macrosystems establish loops that are "closed," at least in part, with various sectors of the ventral mesencephalic dopaminergic neuronal complex. In this regard, the organization is quite reminiscent of the parallel, segregated connectional relationships of the macrosystems with the basal forebrain cholinergic neurons. Thus, it can be postulated, as it was for the basal forebrain cholinergic neurons, that a particular complement of dopaminergic neurons associated with each macrosystem is integral to the particular functional character of that macrosystem and thus might be instructively included as part of its intrinsic circuitry.

5.3.4 Outputs to Thalamus and Epithalamus

The Questions of Open Cortico-Basal Ganglia-Thalamocortical Reentrant Circuits and the Transfer of Information between Circuits

As emphasized in Chapter 3, recognition that the ventral striatopallidal system utilizes a robust, transthalamic reentrant pathway to the prefrontal cortex

[12]Some of the authors reporting on ventral striatopallidal connectivity include Groenewegen and Russchen (1984), Heimer et al. (1991), Somogyi et al., (1991), Groenewegen et al. (1993), Brog et al. (1993), Zahm and Heimer (1993), and Groenewegen et al. (1994). The lateralward spread of accumbens projections into the substantia nigra pars compacta is described by Nauta et al. (1978), Heimer et al. (1991), Zahm and Heimer (1993), Groenewegen et al. (1994), and Haber et al. (1996, 2000).

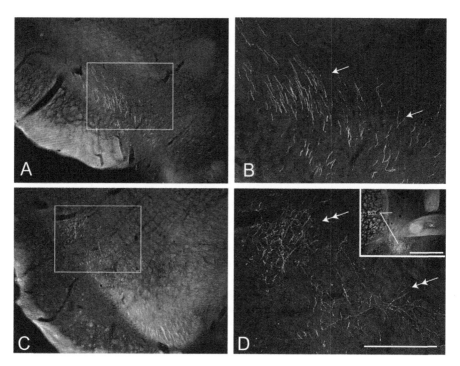

FIGURE 5.6 Sections through the ventral tegmental area (A, B) and rostral mespontine tegmentum (C and D), showing that labeled axonal projections originating in the central division of the extended amygdala pass throught the ventral tegmental area and medial substantia nigra lacking varicosities and thus are likely to be fibers of passage (single arrows). In contrast, these projections exhibit a robust burst of axonal varicosities indicative of functional synapses in the retrorubral field, particularly laterally (double arrows in D). Boxes in A and C are enlarged as B and D, respectively. The inset in D shows the PHA-L injection site in the bed nucleus of stria terminalis (BST) and also shows the robust associational connections within the extended amygdala (indicated by lines). Scale bar in D is 1.0 and 0.4 mm for A, C, and B, D, respectively. Bar in inset is 1.0 mm. Modified from Zahm, 2006, with permission. See color plate.

involving a relay in the mediodorsal nucleus (Heimer and Wilson, 1975; Young et al., 1982) seeded the notion of parallel, segregated cortico-basal ganglia-thalamocortical circuits, or "loops," that figures so prominently in current theories of neuropsychiatric pathophysiology (Lichter and Cummings, 2001). Considering the considerable impact that the perceived "closedness" of these circuits has had on clinical thought, it may be tempting to overstate their closed and segregated nature. Indeed, the open character of these circuits is equally impressive (Zahm and Brog, 1992; Groenewegen and Berendse, 1994; Joel and Weiner, 1994), and several neuroscientists have protested that these circuits incorporate a number of neuroanatomical substrates that should be expected to subserve the likely transfer of information across circuits.

BASIC SCIENCE BOX 7

A Question of Direct Projections

Perhaps basic scientists, and anatomists, particularly, are justly criticized for arguing, seemingly endlessly, what those outside the fray might regard as trivial details. On the other hand, in anatomy, the beauty is in the details. Furthermore, since the goal of neuroscience is not to figure out how problems confronted by brains might be solved but rather to determine how brains solve those problems, a precise appreciation of the structure of a brain must accompany any consideration, and, certainly, any solution, regarding its function. In this book, we have emphasized a role for descending projections from the macrosystems in controlling the activity of ascending neuromodulatory projections. We are far from understanding the nature of this role, and as a consequence, any guidance that we can get from the neuroanatomical organization must be regarded as valuable. It is in this regard that it seems worthwhile to relate some give and take occuring in the field as relates to this issue.

Several investigators (Krettek and Price, 1978; Zahm et al., 1999; Shammah-Lagnado et al., 2001; Gastard et al., 2002) have made note that depositions of anterogradely transported compounds (see Basic Science Box 1, Chapter 2) in central extended amygdala structures, including the central nucleus of the amygdala (CeA) and dorsolateral part of the bed nucleus (BSTdl) of the stria terminalis produce labeled fibers in the ventral tegmental area (VTA) and medial substantia nigra compacta (SNc) that largely lack the varicose "beaded" appearance of axon terminals with synaptic specializations (Fig. 5.6). Such labeled fibers pass through the VTA and medial SNc, showing little evidence of synaptic differentiation until they reach a site in the lateral part of the SNc and suprajacent retrorubral field, where the projection ramifies extensively and exhibits numerous varicosities. This part of the retrorubral field, which was first emphasized by Gonzales and Chesselet (1990) as the main site of CeA terminations in the nigral complex, appears to be the main nigral recipient of a robust, functionally significant (Han et al., 1997; Lee et al., 2005; El-Amamy and Holland, 2007) projection from the extended amygdala.

It is on this backdrop that Fudge and Haber (2000), studying the monkey CeA with anterogradely and retrogradely transported tracers, claimed to have identified a substantial projection from the CeA to the VTA. In contrast, Price and Amaral (1981), evaluating tritiated amino acid injections placed in the monkey CeA, illustrated little labeling in the VTA and a strong projection to the lateral pars compacta and retrorubral field, comparable to that seen in rat. Subsequently, Fudge and Haber (2001), also in the monkey, showed extensive projections to the VTA and substantia nigra pars reticulata and compacta following an injection of BDA into the BSTdl, another extended amygdala structure. A number of technical issues, discussed in Zahm (2006), could be raised to explain these disparate results and, for reasons described following, it seems reasonable to advocate for continuing attention to the problem until a resolution is achieved.

To wit, lack of consensus regarding the existence of a robust, monosynaptic projection from central extended amygdala to the VTA enables a conviction held by behavioral neuroscientists that the pathway should exist (e.g., Parkinson et al., 1999; Hall et al., 2001). Indeed, the existence of such a pathway would fit well with the apparent capacity of drugs or toxins infused into the central nucleus of the amygdala to alter the levels of extracellular dopamine in nucleus accumbens by an action presumed to be mediated by direct projections to the VTA (Louilot, 1985; Simon et al., 1988; Robledo et al., 1996). If, however, the central extended amygdala projection passes through the VTA without synapses, as much of the extant anatomical data indicates, then largely polysynaptic pathways through, for example, the frontal cortex, lateral hypothalamus, and/or mesopontine tegmentum (Nauta and Domesick, 1978; Zahm, 2000; Howland et al., 2001; Zahm et al., 2001; Ahn and Phillips, 2002; Fadel and Deutch, 2002; Phillips et al., 2003) must be considered as substrates for the observed effects of CeA stimulation of dopamine release in the accumbens. To those not in the fray, this may seem a trivial detail. However, possibilities for intrinsic and extrinsic regulation of information arriving across polysynaptic pathways would be expected to far exceed what might be exerted by influences transmitted monosynaptically. Indeed, a recent combined retrograde and anterograde labeling study has demonstrated that the inputs to the VTA originate in numerous cortical and subcortical structures, of which many project at best moderately to the VTA itself, but, in addition, moderately to strongly to one or more other structures that, in turn, innervate the VTA (Geisler and Zahm, 2005). Unfortunately, the challenge to understanding the nature of the regulation of VTA activity is substantially increased by having to consider the properties of such a highly interconnected network. The extended amygdala is just one contributor to the network, and, thus, the possibilities for CeA modulation of VTA activity would appear to be subsumed by the network and subject to tighter regulation than if the CeA simply projected directly to dopamine neurons in the VTA.

Furthermore, it is necessary to acknowledge another recent finding that is difficult to reconcile with the neuroanatomy. Georges and Aston-Jones (2001; 2002; see also Dumont and Williams, 2004) demonstrated a short-latency excitation of the VTA following electrical and chemical stimulation at a single, focal site in the ventrolateral part of the bed nucleus of the stria terminalis. As just noted, the anatomy appears consistent with few, if any, direct projections from the central division of the extended amygdala to the VTA. However, as just noted, great numbers of VTA-projecting neurons form a densely packed column centered on the medial forebrain bundle that invades numerous forebrain structures, including the lateral and medial hypothalamus, sublenticular region, lateral and medial preoptic regions, diagonal band, septal nuclei, ventral pallidum, and ventral parts of the bed nuclei of the stria terminalis (Geisler and Zahm, 2005). Many of the neurons in this column are thought to be glutamatergic—that is, excitatory (McDonald et al., 1989; Takayama and Miura, 1991; Sun and Cassell, 1993; Korotkova et al., 2004; Geisler et al., 2007). Since the structures in this column without exception have direct and indirect projections to the VTA, it is perplexing as to why stimulation of the ventrolateral BST *uniquely* elicits excitatory responses in the VTA. One can readily surmise that we have a lot of ground to cover before we will fully understand structure-function in this part of the brain. However, we do seem to be starting to grasp the scope of the problem.

Haber (2003), for example, has suggested that the massive cortico-thalamic projections are organized such that one would anticipate a substantial feedforward routing of impulse conduction from limbic lobe circuits via executive circuits to cortical motor circuits. A similar idea was put forward in regard to the cortical-ventral basal ganglia-thalamocortical circuits (Zahm and Brog, 1992), and both of these concepts are reminiscent of the notion of a feedforward progression of information flow in the mesotelencephalic dopamine projections that was discussed in a preceding passage. Interestingly, all of these neuroanatomical considerations favor a general flow of information from ventromedial to dorsolateral parts of the striatal complex, consistent with the idea of a passage of information from "motivational" to "motor" parts of the cortico-basal ganglia-thalamocortical circuitry.

Pertinent in this regard are the extraordinary findings, first noted in Chapter 3, of Earl Miller and his colleagues (Pasupathy and Miller, 2005), who have shown experimentally that activity in the cortico-subcortical reentrant circuits may be initiated at a level other than the cortex. They demonstrated that responses associated with light stimulus-evoked visual tracking movements can be recorded in striatal neurons before they are detected in the cortical areas that give rise to corticostriatal projections onto those striatal neurons. This implies that the excitation of the striatal neurons and, by extension, a particular cortico-basal ganglia-thalamocortical loop, can be due to other than cortical input. One possibility for such stimulation to occur, at least in this experimental paradigm, would be via intralaminar thalamostriatal inputs, which themselves may be ignited by visual stimulation of fast pathways to the tectum and associated parts of the reticular formation.

Relays in the Thalamic Midline-Intralaminar Groups and Epithalamus

Albeit with varying strengths, all macrosystems also project to the cortex via synaptic relays in midline-intralaminar and thalamic reticular nuclei, which are usually included within the reticular formation (Ramon-Moliner, 1975) and undoubtedly play a role in regulating cortical arousal. As noted in the preceding paragraph, only dorsal and ventral striatopallidum utilize thalamic relay nuclei—that is, the ventral tier nuclei (ventral anterior and ventral lateral in the primate, ventromedial and ventrolateral in the rat) and the mediodorsal nucleus, respectively. However, although formally categorized as relay motor nuclei (Price, 1995), even the ventral tier nuclei are thought to possess substantial nonspecific character due to their dense, widespread projection to the superficial part of cortical layer I and the substantial inputs that they receive from the lateral habenula, mesopontine and medullary reticular formation, and related structures, such as the parabrachial nucleus. Dorsal and ventral striatopallidum and the septal-preoptic system also project to epithalamus—that is, lateral habenula. This in turn projects to the ventral tier thalamic nuclei, ventral tegmental area, and the dorsal, median, and paramedian raphe formations in the mesopontine tegmentum—that is, to the reticular formation, including the dopaminergic and serotoninergic ascending modulatory systems.[13]

5.4 INTERACTIONS BETWEEN MACROSYSTEMS

5.4.1 Independent Information Processing in Macrosystems

Numerous cortical areas, particularly those comprising the limbic lobe, innervate more than one macrosystem (Figs. 5.2A and B and 5.4C and D), which suggests that macrosystems may utilize similar cortically derived information to achieve different functional ends.[14] This in turn suggests the macrosystems should operate independently, at least in part and thus should be minimally interconnected with each other. This notion fits with ideas expressed in a recent review of theories invoking modular functional-anatomical organization of brain, wherein Záborszky (2002) identified a common theme: that fundamental processing units (modules) operate largely independently of each other to produce distinct outputs for subsequent combination and recombination. Interestingly, this is precisely how we envision the segregated operations of the macrosystems. Indeed, the intuitive appeal of the notion of segregation of processing is tied to the hypothesis that macrosystem outputs ultimately compete and cooperate for access to the motor and cognitive apparatus.

---

[13]Some of the relevant papers here include Nauta and Mehler (1966), Kemp and Powell (1971), Pasquier et al. (1976), Swanson (1976), Herkenham and Nauta (1977, 1979), Herkenham (1979, 1986), Swanson et al. (1984), Groenewegen et al. (1993), Zahm et al. (1996), Geisler et al. (2003), Geisler and Zahm (2005), and Geisler et al. (2007). Of particular interest, a recent study has demonstrated that the lateral habenula is a source of negative reward signals in the dopaminergic mesencephalic ventral tegmental area (Matsumoto and Hikosaka, 2007).

[14]Included among papers relevant to this matter are McGeorge and Faull (1989), Groenewegen et al. (1990, 1997), Berendse et al. (1992), Brog et al. (1993), McIntyre et al. (1996), McDonald et al. (1996, 1999), Shi and Cassell (1998), Zahm (1998, 2000, 2006), and Reynolds and Zahm (2005).

From a neuroanatomical point of view, the input structures are the most segregated components of the macrosystems and thus the most likely substrate for segregated information processing. In accord with this consideration, macrosystem input nuclei project neither back to cortex to the thalamus nor, arguably, to other macrosystems. There is no neuroanatomical evidence, for example, for direct interconnections between the dorsal striatum and any other macrosystem input structure. One might be less confident, however, in proclaiming a segregation of the ventral striatum, particularly as regards connections with the extended amygdala, which some workers claim to have identified and others deny. Because of the technical nature of the arguments on each side of this issue, it is discussed in some detail in Basic Science Box 8.

Likewise, the extent to which lateral septum projects to or receives projections from the shell of the accumbens and bed nucleus of stria terminalis is open to question. These structures physically abut each other, and it is generally difficult to define discrete boundaries separating them. As a consequence, it is not easy to decide whether fibers that cross from one macrosystem to the other within these boundary regions are true projections or simply have breached an artificially designated boundary in "ectopic" fashion. Following infusions of tracer into the bed nucleus, labeled fibers always occupy a narrow fringe of the accumbens shell but only where it abuts the bed nucleus, suggesting that the two structures are indeed related by connections but only at the boundary zone. However, the fibers never penetrate far into the accumbens, which leads one to think that the two are largely not interconnected. Overall, we incline toward the view that the different macrosystems are negligibly interconnected with each other but interact instead through actions exerted by their outputs, which converge at numerous sites in the reticular formation.

5.4.2 Potential Interactions in Lateral Hypothalamus, Ventral Mesencephalon, and Mesopontine Tegmentum

Some level of indirect (polysynaptic) intercommunication between macrosystems might be expected in their rich reciprocal interconnections with the lateral hypothalamus (Zahm, 2000). Indeed, there is considerable apparent overlap of macrosystem outputs—for example, from the accumbens shell and extended amygdala in the lateral hypothalamus. However, close inspection reveals that the terminations of ventral striatopallidal and extended amygdaloid projections occupy largely distinct territories in the lateral hypothalamus (Zahm et al., 1999). This is not to say, however, that these projections are entirely nonoverlapping, and, indeed, foci of substanial overlap of projections likely exist. The septal-preoptic system also gives rise to substantial inputs to the lateral hypothalamus, and although these have not been evaluated with

BASIC SCIENCE BOX 8

Some Examples of Relevant Technical Considerations

The theory described in Chapter 5 of this book predicts independent processing of information by macrosystems, likely at the level of the macrosystem input nuclei—that is, the striatum, including the accumbens, lateral septum, central amygdaloid nucleus, and bed nucleus of stria terminalis. In accord with these considerations, injections of anterogradely transported tritiated amino acids or PHA-L into the central nucleus of the amygdala, a prominent extended amygdala input structure, produce no labeling in the accumbens, a ventral striatopallidal input structure (Krettek and Price, 1978 [cat, rat]; Price and Amaral, 1981 [monkey]; Zahm et al., 1999 [rat]). Nonetheless, Fudge et al. (2002) concluded that the central nucleus of the amygdala (CeA) in the old world monkey projects robustly to the shell of the nucleus accumbens. However, of several accumbens injections of retrogradely transported tracer considered in Fudge et al. (2002), the only one that resulted in significant retrograde labeling in the CeA was close enough to the bed nucleus of the stria terminalis (BST) that one would anticipate BST involvement at the injection site, which would explain the retrograde labeling in the CeA, insofar as the CeA is well known to project strongly to the BST. Furthermore, the medial amygdala also was labeled by this injection but should have remained unlabeled unless the injection involved the BST, since the medial amygdala does not project to the accumbens (Canteras et al., 1995).

As regards the anterograde tracing data shown by Fudge et al. (2002), the studies were done with so-called "bi-directional" tracers—tracers that are transported anterogradely *and* retrogradely. Bidirectionally transported tracers typically produce massive backfilling of axons, which then transport tracer anterogradely in axon collaterals (Chen and Aston-Jones, 1998), thus producing a misleading result. Such injections into the rat CeA result in strong filling of neurons in the basal and accessory basal complexes of the amygdala, which also project quite strongly to the accumbens (Krettek and Price, 1978), thus giving the erroneous impression of a labeled CeA-accumbens projection (Fig. A). As just noted, PHA-L, an axonal tracer that is transported only anterogradely, does not produce anterograde labeling in the accumbens following injections in the CeA (Fig. B). Consistent with these observations, a number of figures in Fudge et al. (2002) show anterograde label in parts of the ventral striatum where injections of retrograde tracer, illustrated in the same study, produced no retrograde labeling in the CeA.

As regards direct projections in the reverse direction—that is, from the accumbens to the extended amygdala, published tract-tracing data provide little support for them in the rodent (Nauta et al., 1978; Zahm and Heimer, 1990, 1993; Heimer et al., 1991; Usuda et al., 1998) or cat (Groenewegen and Russchen, 1984). By comparison to the projections from the accumben shell to its main targets, such as the ventral pallidum, lateral hypothalamus, and ventral mesencephalon, negligible numbers of labeled fibers have been observed in the central nucleus of the amygdala (Heimer et al., 1991) or bed nucleus of stria terminalis (Usuda et al., 1998) following shell injections of anterogradely transported substances. Indeed, the few reported may simply reflect involvement at the injection site of extended amygdala structures, such as the BST, which physically abuts the caudomedial shell.

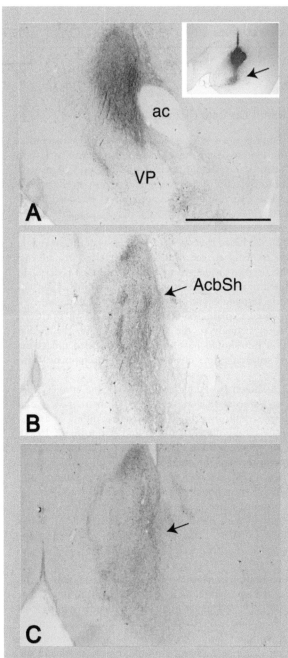

BSBOX 8 FIGURE A Micrographs illustrating axonal transport following an injection of bidirectionally transported cholera toxin β subunit (CTβ) into the central nucleus of the amygdala (CeA, the injection site is shown in the inset in A). In contrast to PHA-L injections into the CeA, which produce negligible anterograde labeling in the nucleus accumbens shell (AcbSh, see BSB 8 Fig. B), this injection produced robust labeling there (arrows in panels B and C; C is 500 μm rostral to B). CTβ is transported avidly in both the retrograde and anterograde directions, however, and in this case there is robust retrograde labeling of neurons in the accessory basal nucleus of the amgdala (AB, arrow in inset in panel A) and scattered labeling throughout the basal-accessory basal complex of the amygdala. The AB and other parts of the basal complex project robustly to the accumbens shell (Krettek and Price, 1978; Brog et al., 1993). Thus, accessory basal neurons transported CTβ retrogradely from the injection site in the CeA and then anterogradely to the AcbSh. Additional abbreviations: ac—anterior commissure; VP—ventral pallidum. (Reprinted from Zahm, 2006, with permission.) See color plate.

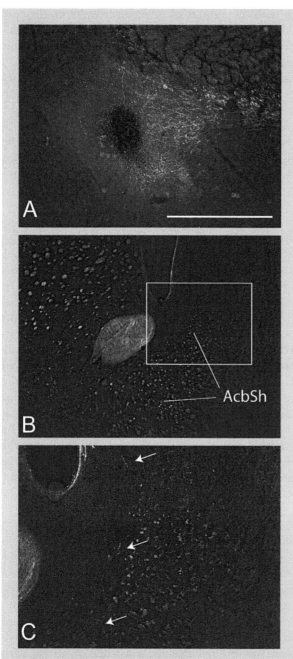

BSBOX 8 FIGURE B Micrographs showing sections illuminated in dark-field mode from a case in which a PHA-L injection was made in the central nucleus of the amygdala (CeA, shown in panel A). Panels B and C illustrate the accumbens shell (AcbSh) where negligible anterograde labeling (arrows) is observed. The box in B is enlarged in C. Scale bar: 1 mm in A, 0.4 mm in C. (Modified from Zahm, 2006, with permission.) See color plate.

regard to potential overlap with the outputs of other macrosystems, they are sufficiently extensive that the possibility of crosstalk in the lateral hypothalamus between septal-preoptic and other macrosystems surely also exists. This complicated neuroanatomical organization clearly begs additional study.

Macrosystem interactions also could be mediated by ascending modulatory projections. For example, ventral mesencephalic dopamine neurons project to basal forebrain cholinergic neurons and are themselves strongly innervated by cholinergic projections from the PPTg. The functional impact associated with the innervation of one ascending modulatory system by another is likely to be substantial and far reaching. One can envision virtually limitless permutations involving interactions between the macrosystems and ascending modulatory pathways.

5.4.3 A Potentially Robust Transthalamic Feed-Forward Relationship

Extended amgydala outputs project abundantly to the paraventricular nucleus of the thalamus (Grove, 1988; Arluison and Derer, 1993; Dong et al., 2001), which in turn provides a very dense projection to the shell of the nucleus accumbens (Bubser et al., 2000). This pathway could convey a rather robust influence from extended amygdala to ventral striatopallidum that, interestingly, is not apparently reciprocated by a correspondingly robust pathway from striatopallidum via the thalamus to extended amygdala. Data conveyed by the extended amygdala-paraventricular thalamus link could substantially influence the processing of information by the ventral striatopallidum. This type of feed-forward transfer of information has been proposed to occur *within* but not previously *between* different functional-anatomical systems. As just noted, a transpallidothalamic-corticostriatal pathway from the shell to the core of the nucleus accumbens has been described, as has a substrate for the passage of neural information lateralward in the striatum via ventral striatonigral projections to more laterally projecting dorsal nigrostriatal neurons.

5.5 THE PROBLEMS OF SUBDIVISION, BOUNDARIES, AND THE REDUCTIONIST APPROACH

The observation that subsets of basal forebrain, cortical, diencephalic, and brainstem structures segregate into more or less distinct, mutually interconnected constellations that exhibit highly characteristic patterns of intrinsic and extrinsic connections gave rise to the concept of basal forebrain functional-anatomical systems. Thus, it was posited that the existence of several such systems centered on, for example, striatopallidum, extended amygdala, and septum. Swanson (2000) has proposed another such system related to the mammillary body. Here, we have addressed the downstream and reentrant connectional relationships of such systems, as well as their connections to each other, and it has been pointed out that details of this connectivity are likely to be critical with respect to how these systems contribute to brain function.

Although the discussion has focused on "macrosystems," such as ventral striatopallidum, extended amygdala, and the septal-preoptic system, it may be that these major basal forebrain functional-anatomical systems are subject to functionally relevant subdivision. The striatal complex, for example, comprises dorsal and ventral territories, which already are regarded as distinct input nuclei representing distinct macrosystems—dorsal and ventral striatopallidum. Dorsal and ventral striatum, however, can be further subdivided into subterritories, as in the core, shell, and rostral pole of the accumbens or quadrants of the caudate-putamen (Zahm and Brog, 1992; Heimer et al., 1995; Grimm et al., 2001), of which all possess unique character in terms of the fine grain of their cytoarchitectural, molecular biological, and neurochemical composition, and the specific identities of afferent and efferent connections. Subterritories can be further resolved into connectionally and neurochemically distinct subdivisions. The accumbens shell, for example, appears to comprise multiple subdivisions (Zahm et al., 1998; Todtenkopf et al., 2004), which, in concert with interconnected districts in cortex, diencephalon, and brainstem, may operate as functional units quite along lines proposed herein. Furthermore, it should be emphasized that distinct anatomical boundaries separating such putative functional units may not be apparent. Just as transition zones of ambiguous neuroanatomical character insinuate between major functional-anatomical entities, such as striatopallidum and extended amygdala (Alheid et al., 1994), lesser subdivisions as proposed here may well merge with each other across transitional zones rather than boundaries. However, it is also possible that subdivisions of functional-anatomical macrosystems work together to accomplish specific tasks. Essential objectives of future work must be to determine the minimal makeup of functionally potent macrosystems and define their functions. For the present, it is important to emphasize that the systems discussed in this book are neuroanatomical constructs. The manner in which they contribute to function remains to be understood.

Since Heimer and Wilson's (1975) benchmark paper describing striatopallidum paved the way for the conceptualization of forebrain functional-anatomical systems, some investigators have suggested that striatopallidum represents a prototypical functional-anatomical system, even to the extent of advocating that all such systems are striatopallidum or specializations of it (Cassell, 1998; Swanson and Petrovich, 1998; Cassell et al., 1999; Swanson, 2000, 2003). While acknowledging the likely validity of this conservative (reductionist) viewpoint, consistent with the generalized template proposed by Alheid and Heimer (1988), it can also be pointed out that neurochemical composition, receptors, signaling pathways, peptides, populations of associated interneurons, and intrinsic and extrinsic connectional relationships amply distinguish different functional-anatomical systems such that each should be expected to process information in a distinct way. Thus, the unique functional-anatomical identities of such systems are critical to their hypothesized capacity to generate distinct patterns of output.

5.6 SOME FUNCTIONAL CONSIDERATIONS AND CONCLUDING COMMENTS

Insofar as we think of the macrosystems as taking over some of the functions that have been attributed to the limbic system, it certainly would be satisfying if we could associate macrosystem actions with emotions such as anger, envy, jealousy, and the like. But perhaps the complexities of florid emotional states such as these are sure to defeat any attempt to marshal our nascent appreciation of the macrosystems in explanation. Furthermore, it may be that some of what we think of as emotions, such as underlie, for example, rage, attack, and some stereotyped sexual posturings are programmed into the diencephalic and lower brainstem circuitry subject to disinhibitory release dependent on the actions of descending projections (Swanson, 1987; Keay and Bandler, 2001), perhaps, at least in part, by the actions of macrosystems. As was discussed in an earlier part of the book, rats deprived of the entire forebrain, including even the diencephalon, exhibit attack behaviors (Woods, 1963). This, however, is not to say that we might not profitably consider macrosystem contributions to affect, as they relate to less dramatic aspects of emotional life, such as, for example, the process of deciding.

Although life situations tend to pit the consummation of appetitive, affiliative and/or sexual drives, or combinations of these, against combinations of near or distant and varyingly serious threats, imminent threat to life occupies but a fraction of most organisms' cognitive lives. Mostly, creatures' waking hours are consumed in varyingly intense vigilance punctuated by more or less routine risk-benefit analyses aimed at determining if potentially executable actions should be taken or avoided. So goes the continual quest to satisfy physical and psychical wants in ever-changing, varyingly dangerous circumstances. In view of the tremendous adaptive value of competent risk-benefit analyses, a comprehensive understanding of their neural basis is an essential objective. Impaired risk-benefit assessment, which results in maladaptive choices, interacts with a broad range of human disorders of affect, from chronic anxiety and depression to fear and panic states (Hotz and Helm-Estabrook, 1995; Davey and Levy, 1999; Rapport et al., 2001), and its fallout likely spans the full gamut of human social problems, from joblessness and ruinous child rearing to various addictions (Beatty et al., 1993). Indeed, addiction is defined operationally as a cluster of compulsive activities that are intentionally repeated despite the explicit knowledge that gravely adverse consequences inevitably follow (Volkow and Li, 2004)[15]. From this viewpoint, at least part of the cycle of addiction can be regarded as a breakdown in risk-benefit assessment. In view of these considerations, it is important to understand not only how the brain processes reward and risk but also how reward and risk interact in the synthesis of adaptive behavior and how the interaction might be changed by neuropathology, psychopathology, and adaptive neuroplasticity associated with drug use (see also Clinical Box 15).

A Cortical Substrate of Risk-Benefit Analysis?

An argument might be made that brain circuitry capable of subserving risk-benefit assessment is well enough reflected in the disinhibited, impulsive, perseverative behavioral syndrome associated with damage to the ventromedial prefrontal cortex, as in, for example, patients described by Damasio (1994). However, ventromedial prefrontal cortex lesions are rare, whereas impaired risk-benefit assessment is ubiquitous in a variety of human neurological and behavioral disorders (Frankel, 1978; Ridley, 1994; Joseph, 1999)—a fact that encourages consideration of other substrates. Indeed, it can be difficult to distinguish clinically between cortical and subcortical components of neurological deficit insofar as cortical damage in itself severely alters the functional potency of subcortical targets, due to the loss of cortico-subcortical connections (e.g., Eslinger and Grattan, 1993; Adair et al., 1996; Damasio, 1998;

---

[15]We have hesitated to discuss in any detail the important areas of therapeutically administered psychotropic drugs, drugs of abuse and drug dependency in this book because, despite the existence of a wealth of literature addressing these topics, implications attached to the diverse actions of such drugs on multiple macrosystems have received little attention from research scientists so far. Consider that psychostimulant drugs act by precipitating the release and/or blocking the re-uptake of catecholamines, including dopamine, norepmephrine and epinephrine, and the indoleamine, serotonin, and that ascending neuromodulatory projections containing these neurotransmitters richly innervate all of the macrosystems. Consider further that the receptor binding, uptake mechanisms and pre- and postsynaptic actions of these transmitters differ across the normal macrosystems and thus the macrosystems would be expected to respond differently to drug actions. Opiate receptors, the targets of heroin and morphine, also are widely distributed through the macrosystems and subject to similar considerations. Consider that repeated administrations of both psychostimulant and opiate drugs can induce varyingly permanent neuroadaptive alterations in cellular mechanisms and signaling that, given the preceding considerations, are likely to differ in the different macrosystems. Finally, consider that macrosystem outputs are strategically disposed to influence cognitive, emotional, visceromotor and somatomotor expression and that drug induced changes in the functional potencies of the macrosystems thus would be expected to be expressed in each of these domains. As regards the therapeutic use of psychotropic drugs, which typically target monoamine or indoleamine systems, similar considerations with regard to the macrosystems would seem to apply. Since these drugs are frequently administered to children and adolescents, to the panoply of possibilities described in the preceding must be added concerns regarding cascading disruptions of normal developmental processes. Adequate consideration of these topics signifies a long road ahead.

CLINICAL BOX 15

Disorders of Motivation

Traditionally, behavioral pathologies are placed in three main groups—namely, disorders of conation, cognition, and emotion. However, essential to behavior, and not so easily placed in one of these categories, is action, in the sense of imparting direction and change to the status quo. Included would be impulse, desire, and striving toward goals, the underlying individual motivations being variously available (or not) to consciousness. Disorders of motivation seem to have been curiously neglected in psychiatric practice, yet patients with primary amotivational states, referred to as abulia, are frequently encountered, and secondary disorders of motivation are noted in a wide variety of psychopathologies, characterized by apathy, flattened affect, and anergia (as seen in the chronic fatigue syndrome or in hypothyroidism).

Disorders of motivation are seen in schizophrenia (negative symptoms), depression (apathy, anhedonia), and in the emotional indifference of a Klüver-Bucy syndrome. Increased states of motivation occur in mania and also are a manifestation of addictions. The intensely driven motor behavior of Gilles de la Tourette's syndrome may be seen as reflection of a hypermotivational state, and, since many such patients have a comorbid obsessive-compulsive disorder, this syndrome may also be viewed in part as a dysregulation of motivation.

Abulia, in which the patient literally has no will, is a primary disorder of motivation. Abulic patients will sit in a chair all day long and do nothing. Whereas in apathy patients simply can't be bothered to do things, in abulia there is simply no desire. Further, when asked if they ever feel bored, patients with abulia simply say no. Abulia is frequently encountered in patients with basal ganglia disorders, especially those involving damage to the connections between the anterior cingulate area and striatum. Abulia may seen in Parkinson's disease but also following strokes involving the basal ganglia and head injuries with damage to fronto-striatal structures. Thus, the limbic forebrain and dopamine therein have such important consequences for driving behavior that damage to them may be expected to have profound clinical effects. Indeed, they do, and yet, disorders of motivation find no place in such diagnostic manuals as the DSM-IV and so are poorly recognized clinically and consequently underdiagnosed. The unraveling of the neuroanatomy of the limbic forebrain allows for an understanding of these behaviors, which correlate closely with that prime human force, the will, once regarded as an independent psychological faculty, now viewed as tightly bound with identifiable neuroanatomical and neurochemical substrates.

Damasio et al., 2000). In this regard, for example, the effect of basal amygdaloid lesions on decision making is reported to approximate that of ventromedial prefrontal cortex lesions (Bechara et al., 1999), pointing more toward subcortical structures, to which both project as the operative substrates (see also Clinical Box 16).

CLINICAL BOX 16

Subcortical Versus Cortical Psychopathologies

The terms *cortical* and *subcortical* are frequently used in this book to distinguish the cerebral cortex from all in the brain that is not cortical, particularly the telencephalic nuclei and related brainstem structures. Thus, it would not be surprising if the discrete boundary separating the cortex from the rest of the brain would be associated with a watershed defining different neurological syndromes. Indeed, the idea emerged a number of years ago that dementia syndromes with pathology primarily in subcortical structures may have a clinical profile that differs predictably from the cortical dementias. The paradigmatic example of a cortical dementia is Alzheimer's disease, with a behavior profile that includes aphasia and apraxia, seen as largely due to loss of neuronal tissue in the isocortex.

The first descriptions of subcortical dementia came from Albert et al. (1974). The presentations were flavored by changes of personality, emotionality, and cognition, reflecting reductions in behavioral flexibility, inability to manipulate acquired knowledge, and slowness of information processing. The subcortical picture was seen as accompanying such syndromes as Huntington's chorea, Parkinson's disease, hydrocephalus, and head injuries and to be associated principally with damage to the basal ganglia. On neurocognitive testing such patients exhibited executive function deficits, but importantly, and especially early on in the disorder, they did not show episodic memory deficits or other cortical signs, such as topographical disorientation. Thus, a deterioration of cognition and mood disturbances distinguished these syndromes. In psychiatry, the cognitive changes of schizophrenia were thought representative of a subcortical syndrome, as were the more severe cognitive problems of depression.

Although the concept of distinguishing subcortical and cortical dementias has been advocated strongly by some (Cummings, 1990), it has not gained universal acceptance. This is in part because it has become clear, as neuropathological studies have progressed, that the classical cortical dementias are accompanied by widespread pathology beyond the isocortex (for example, the entorhinal and hippocampal changes of Alzheimer's disease). But, in addition, the subcortical-cortical breakdown has been superseded by a major alternative classification, which has gained greater clinical currency—namely, to distinguish Alzheimer's disease (AD) from the fronto-temporal dementias. AD presents classically with failing memory, irritability, and language disturbances but good preservation of the personality. In contrast, frontotemporal dementia begins with abnormalities of behavior, emotional changes, and personality deviations. Differences in the underlying pathologies have been shown, with amyloid plaques and neurofibrillary tangles being the hallmark of AD, amyloid being a crucial protein in the development of AD. In contrast the pathology of the frontotemporal dementias reveals more gliosis and spongiform changes. However, even these distinctions are now fading in the shadow of newer classifications based in genetics and the structures of specific proteins, such as tau. Abnormalities of tau protein are found in several neurodegenerative disorders, both sporadic and inherited, suggesting that the tauopathies associated with mutations of the tau gene can independently lead to neurodegeneration.

If anything, this book has championed the point of view that any split between cortical and subcortical structures is a poor guide to understanding behavior and its regulation. The concept of the rich and intimate interplay between the cortex, particularly that of the greater limbic lobe, and the basal forebrain macrosystems, mediated by cortico-striato-pallido-thalamic reentrant circuits, ascending "state-setting" projections and macrosystem outputs to motor effectors reveals the essential inseparability of the cortical from the subcortical. In clinical settings, the ensuing symptomatology inevitably reflects this dynamic.

Subcortical Substrates of Risk-Benefit Analysis?

The hypothesis that the neural correlates of affect are generated subcortically is consistent not only with a wealth of experimental data, of which only a small fraction will be referred to later in this section, but also with an understanding of the fundamental organization of vertebrate brain along lines described in this book. From this point of view, subcortical information processing systems are substrates within which outputs from all parts of cortex are processed prior to engaging multisynaptic pathways to (1) hypothalamic and brainstem somatomotor and visceromotor effectors and (2) back to the cortical cognitive apparatus via "reentrant" trajectories. As has been thoroughly documented in the preceding chapters, the striatopallidal complex, extended amygdala, and septum-preoptic system are apt examples of this kind of brain organization, each emulating organizational characteristics and patterns of connections that, while fundamentally striatopallidal, are nonetheless highly distinct. It is well established that each of these functional-anatomical constructs is innervated by the same high-level associational areas comprising the greater limbic lobe, including the prefrontal cortex, basolateral amygdala, and ventral hippocampus, as well as by dopaminergic and "nonspecific" thalamic projections. Furthermore, we have described here how each relates preferentially to specific parts of the cholinergic and dopaminergic ascending neuromodulatory projection systems. These features lead to the hypothesis that each such system is capable of evaluating ongoing cortical representations and discriminating specific affect-related aspects of the environmental and internal circumstances they reflect. In turn, the subcortical system outputs are conveyed via thalamocortical and ascending and descending modulatory pathways to compete and cooperate for representation in ensuing syntheses of cognitive, somatomotor, and visceromotor components of behavior (Zahm, 2006).

Abundant evidence indicates that subcortical systems, including ventral striatopallidum, extended amygdala, and the septal-preoptic system, are essential to the integration of affective and locomotor components of responses—for example, to novel objects, the environment, appetitive and aversive stimuli, fear-arousing stimuli, discrete and contextual anxiogenic stimuli, and stimuli associated with prior drug administration.[16] Indeed, affect largely dictates what the motor components of behavioral responses to stimuli will be: approach, avoidance, attack, flight, or freezing. The ventral striatopallidum, particularly its accumbens part, is associated with locomotor and incentive-reward effects that are among the most robust elicitable from any brain structure (e.g., Kelly et al., 1975). The core subterritory of the accumbens, in particular, gates transitions between motivation and action initiation, such as in conditioned orienting to appetitive stimuli, Pavlovian approach (Parkinson et al., 1999; Cardinal et al., 2002), and instrumental responding (Kelley et al., 1997). The accumbens shell, in contrast, appears to modulate the vigor of behavioral responses (Cardinal et al., 2002; Salamone and Correa, 2002; Kelley, 2004). Indeed, much evidence indicates that approach (or its absence) is controlled by the accumbens, apparently by mechanisms largely involving disinhibition. Thus, circumstances that produce generalized reductions in neuronal activity in the accumbens lead to generalized increases in locomotor activation state. Locomotion is robustly increased by administering GABA (Shreve and Uretsky, 1988) and dopamine (Jones and Mogenson, 1980; Pijnenburg and Van Rossum, 1973) receptor agonists, stimulants of dopamine release (Kelly et al., 1975), and antagonists of the dopamine transporter (Boulay et al., 1996) to the accumbens. Similarly, bilateral cell-depleting lesions of the accumbens increase locomotion (Woodruff et al., 1976; Koob et al., 1981; Kelly and Roberts, 1983).

Alternatively, extended amygdaloid structures, such as the central nucleus of the amygdala and bed nucleus of the stria terminalis (Alheid and Heimer, 1988), regulate the affective quality and magnitude of an organism's behavioral responses and related autonomic responses (e.g., LeDoux et al., 1988) to internal and environmental sensorimotor cues, most particularly as relates to the fear-arousing content of the stimulus (LeDoux et al., 1988; Gallagher et al., 1990; Davis et al., 1993; Gallagher and Holland, 1994; Robledo et al., 1996; Davis and Shi, 1999). Thus, affective, locomotor, and visceromotor components of behavioral freezing and fear potentiation of startle are disrupted by lesions or inactivations of the extended amygdala (e.g., Davis et al.,

[16]Relevant literature includes Drugan et al. (1986), Rebec et al. (1997a, b), Davis and Shi (1999), Parkinson et al. (1999), Degroot and Parent (2001), Degroot et al. (2001), Reynolds and Berridge (2001), Cardinal et al. (2002a), McFarland et al. (2003, 2004), Kelley (2004), and Sheehan et al. (2004).

1993; Davis and Shi, 1999), which indicates that extended amygdala is concerned in a major way with threat detection.

Hippocampus and septum are implicated in avoidance retention, anxiety, and conflict and anticonflict behaviors (Drugan et al., 1986; Degroot and Parent, 2001; Degroot et al., 2001). A recent large literature review (Sheehan et al., 2004) supports a critical role for the lateral septum in the regulation of neural processes related to mood and motivation, implicating it in a broad and diverse range of behaviors and psychopathologies, including bimodal effects on fear responding, behavioral manifestations of depression and psychosis, and the therapeutic actions of antidepressant and antipsychotic drugs.

How functional-anatomical systems might interact in behavioral synthesis might be illustrated by the undoubtedly grossly oversimplified example of a bright, open space in the center of which lies a palatable morsel visible to a hungry rat. Extended amygdala might extract information from the cortical representation of this particular set of stimuli about threat potential related to the light and openness, and thus encourage reluctance to approach and consume. The same or related data shunted through the nucleus accumbens may be assessed for positive motivational—that is, approach, value, impelling the rodent to get the food. Thus, the opposing influences are weighed in the ensuing synthesis of behavior. The actual manner in which macrosystems operate is undoubtedly not so simple, however, not least because of the significant redundance in their involvements in behaviors. Responses to or effects upon fear and anxiety have been identified in the extended amygdala (e.g., Davis and Shi, 1999) lateral septum (Drugan, 1986; Degroot and Parent, 2001; Degroot et al., 2001), and accumbens (Reynolds and Berridge, 2001). Pavlovian approach and responses to conditioned appetitive stimuli appear controlled by the accumbens (Parkinson et al., 1999; Kelley, 2004) and extended amygdala (Han et al., 1997; Cardinal et al., 2002). Each of these systems can elicit or suppress locomotion (Cardinal et al., 2002; Kelley, 2004; Sheehan et al., 2004). Thus, a given emotional stimulus is likely to elicit complex, distinct responses in each of the described functional-anatomical systems, with different consequences in terms of what information each system in turn conveys to the cognitive and motor apparatus.

In terms perhaps more relevant to the everyday lives of people, multiple information processing channels would be expected to increase the breadth and flexibility of cognitive and motor function (see, e.g., Holstege, 1989, 1992; Nieuwenhuys, 1996 [particularly Fig. 8]; but also Alheid and Heimer, 1996). The variegation and redundancy that this circuitry adds to cognitive and motor capacity may, at least in part, underlie the fluid interplay that characterizes the largely graceful way individuals respond to the ever-changing significance and patterning of consecutive and superimposed stimuli: How it is that one moves quite differently to music than silence; reacts more vigorously to unexpected stimuli when nervous; smiles more genuinely on the approach of a friend as compared to for the camera; tosses a balled paper accurately into the can on a first attempt but has difficulty repeating when consciously "aiming." This circuitry may explain how the mask-like countenance of the parkinsonian patient may break into a grin upon the sincere perception of humor.

Concluding Comments

We acknowledge that macrosystems at present remain a theoretical construct, no matter how well they seem to be anchored in what appears to be neuro-anatomical and functional reality. Consequently, there may remain a significant component of speculation in the respective roles we attribute to the basal forebrain macrosystems in regulating behavior. Nevertheless, having examined the available data carefully, one is struck by the likelihood that these systems coexist with the rest of the brain, particularly the greater limbic lobe, above, and brainstem-spinal motor effectors, below, in a state of dynamic equilibrium, to impart affective tone—feelings—to the recollection of past events, acquisition of new stimuli, and execution of adaptive behaviors.

LITERATURE CITED

Adair JC, Williamson DJG, Schwartz RL, Heilman KM. 1996. Ventral tegmental area injury and frontal lobe disorder. Neurology 46:842–843.

Adams F. 1939. The genuine works of Hippocrates. Baltimore: Williams and Wilkins.

Adey WR, Tokizane T. 1967. Structure and function of the limbic system. Prog Brain Res 27.

Adolphs R, Tranel D, Damasio H, Damasio AR. 1994. Impaired recognition of emotion in facial expressions following bilateral damage to the human amygdala. Nature, 372:669–672.

Adolphs RD, Tranel D, Hamann S, Young AW, Calder AJ, Phelps EA, Anderson A, Lee GP, Damasio AR. 1999. Recognition of facial emotion in nine subjects with bilateral amygdala damage, Neuropsychologia 37:1111–1117.

Alexander GE, DeLong MR, Strick PL. 1986. Parallel organization of functionally segregated circuits linking basal ganglia and cortex. Annu Rev Neurosci 9: 357–381.

Alheid GF. 2003. Extended amygdala and basal forebrain. In Shinnick-Gallagher P, Pitkänen A, Shekhar A, Cahill C, editors. The amygdala in brain function. New York: New York Academy of Sciences, pp. 185–205.

Alheid GF, Beltramino C, Braun A, Miselis RR, François C, de Olmos JS. Transition areas of the striatopallidal system and extended amygdala in the rat and primate: Observations from histochemistry and experiments with mono- and transsynaptic tracer. In: Percheron G, McKenzie JS, Féger J, editors. The Basal Ganglia IV, Vol. 41, New Ideas and Data on Structure and Function. New York: Plenum Press, 1994, pp. 95–107.

Alheid GF, Beltramino CA, de Olmos JS, Forbes MS, Swanson DJ, Heimer L. 1998. The neuronal organization of the supracapsular part of the stria terminalis in the rat: the dorsal component of extended amygdala. Neuroscience 84:967–996.

Alheid GF, Heimer L. 1988. New perspectives in basal forebrain organization of special relevance for neuropsychiatric disorders: the striatopallidal, amygdaloid, and corticopetal components of substantia innominata. Neuroscience 27:1–39.

Alheid GF, Heimer L. 1996. Theories of basal forebrain organization and the "emotional motor system." In: Holstege G, Bandler R, Saper CB, editors. The Emotional Motor System. Prog Brain Res 107:461–484.

Alheid GF, Heimer L, Switzer RC III. Basal ganglia. In: Paxinos G, editor. The Human Nervous System. Sydney: Academic Press, 1990, pp. 483–582.

Alheid GF, de Olmos JS, Beltramino CA. 1995. Amygdala and extended amygdala. In: Paxinos G (Ed.) The Rat Nervous System, second ed. San Diego, Academic Press, Inc., pp. 495–578.

Allison AC. 1954. The secondary olfactory areas in the human brain. J Anat 88: 481–488.

Albert ML, Feldman RG, Willis AL. 1974. The "subcortical dementia" of progressive nuclear palsy. J Neurol Neurosurg Psychiatr 37:121–130.

Altschuler LL, Bartzokis G, Grieder T, et al. 1998. Amygdala enlargement in bipolar disorder and hippocampal reduction in schizophrenia: An MRI study demonstrating neuroanatomic specificity. Arch Gen Psychiatr 55:663–664.

Alvarez-Bolado G, Rosenfeld MG, Swanson LW. 1995. Model of forebrain regionalization based on spatiotemporal patterns of POU-III homeobox gene expression, birthdates, and morphological features. J Comp Neurol 355:237–295.

Amaral DG, Avendano C, Benoit R. 1989. Distribution of somatostatin-like immunoreactivity in the monkey amygdala. J Comp Neurol 284:294–313.

Amaral DG, Behniea H, Kelly JL. 2003. Topographic organization of projections from the amygdala to the visual cortex in the macaque monkey. Neuroscience, 118: 1099–1120.

Amaral DG, Insausti R, 1990. Hippocampal formation. In: Paxinos G (Ed.) The Human Nervous System, Academic Press, San Diego.

Amaral DG, Insausti R, Cowan WM. 1987. The entorhinal cortex of the monkey: I. Cytoarchitectonic organization. J Comp Neurol 264:326–355.

Amaral DG, Price JL, Pitkanen A, Carmichael ST. 1992. Anatomical organization of the primate amygdala complex. In: Aggleton JP (Ed.) The Amygdala, Wiley-Liss, New York, 1–66.

Andén NE, Carlsson A, Dahlström A, Fuxe K, Hillarp NÅ, Larsson K. 1964. Demonstration and mapping out of nigro-neostriatal dopaminergic neurons. Life Sci 3:523–530.

Andén NE, Dahlström A, Fuxe K, Larsson K. 1965. Mapping out of catecholaminergic and 5-hydroxtryptamine neurons innervating the telencephalon and diencephalon. Life Sci 4:1275–1279.

Andén NE, Dahlström A, Fuxe K, Larsson K, Olson L, Ungerstedt U. 1966a. Ascending monoamine neurons to the telencephalon and diencephalon. Acta Physiol. Scand. 67:313–326.

Andén NE, Fuxe K, Hamberger B, Hökfelt T. 1966b. A quantitative study on the nigro-striatal dopamine neuron system in the rat. Acta Physiol Scand 67:306–312.

Andreasen NC, Paradiso S, O'Leary DS. 1998. "Cognitive Dysmetria" as an integrative theory of schizophrenia: a dysfunction in cortical-subcortical-cerebellar circuitry? Schizo Bull 24:203–218.

Anthoney TR. 1994. Neuroanatomic and neurologic exam. Boca Raton: CRC Press.

Arendt T, Marcova L, Bigl V, Brückner MK. 1995. Dendritic reorganization in the basal forebrain under degenerative conditions and its defects in Alzheimer's disease.

I. Dendritic ortanization of the normal human basal forebrain. J Comp Neurol 169–188.

Arluison M, Derer P. 1993. Forebrain connections of the rat paraventricular thalamic nucleus as demonstrated using the carbocyanide dye DiI. Neurobiol 337–350.

Arnold SE, Hyman BT, Flory J, Damasio AR, Van Hoesen GW. 1991. The topographic and neuroanatomical distribution of neurofibrillary tangles and neuritic plaques in the cerebral cortex of patients with Alzheimer's disease. Cereb Cortex 1:103–116.

Arnold SE, Trojanowski JQ. 1996. Recent advances in defining the neuro-pathology of schizophrenia. Acta Neuropathol 92:217–231.

Auer DP, Putz B, Kraft E, et al. 2000. Reduced glutamate in the anterior cingulate cortex in depression: an in vivo proton magnetic resonance spectroscopy study. Biological Psychiatry 47:305–311.

Barbeau A, Sourkes TL, Murphy GF. 1962. Les catécholamines dans la maladie de Parkionson. In: Ajuriaguerra J (ed.) Monamines et Systeme Nerveux Central. Symposium Bel-Aire, Geneva, 1961, pp. 247–262. Georg et Co., Geneva.

Bard P. 1928. A diencephalic mechanism for expression of rage with with special reference to the sympathetic nervous system. Am J Physiol 84:490–513.

Bard P, Rioch DM. 1937. A study of four cats deprived of neocortex and additional portions of the forebrain. Johns Hopkins Med J 60:73–153.

Beatty WW, Katzung VM, Nixon SJ, Moreland VJ. 1993. Problem-solving deficits in alcoholics: Evidence from the California Card Sorting Test. J Stud Alcohol 54:687–692.

Bechara A, Damasio H, Damasio AR, Lee GP. 1999. Different contributions of the human amygdala and ventromedial prefrontal cortex to decision-making. J Neurosci 19:5473–5481.

Bechterew V. 1900. Demonstration eines Gehirn mit zerstörung der vorderen and inneren Theile der Hirnrinde beider Schläfenlappen. Neurologisches Centralblatt 19:990–991.

Bench CJ, Friston KJ, Brown RG, et al. 1992. The anatomy of melancholia—focal abnormalities of CBF in major depression. Psychological Medicine, 22:607–615.

Bennes FM. 1993. Neurobiological investigations in cingulate cortex in schizophrenic brain. Schizophrenia Bulletin, 19:537–549.

Berendse HW, Galis-de Graaf Y, Groenewegen HJ. 1992. Topographical organization and relationship with ventral striatal compartments of prefrontal corticostriatal projections in the rat. J Comp Neurol 316:314–347.

Bernier PJ, Parent A. 1998. Bcl-2 protein as a marker of neuronal immaturity in postnatal primate brain. J Neurosci 18:2486–2497.

Biber MP, Kneisley LW, LaVail JH. 1978. Cortical neurons projecting to the cervical and lumbar enlargements of the spinal cord in young and adult rhesus monkeys. Exp Neurol 59:492–508.

Bigl V, Woolf WJ, Butcher LL. 1982. Cholinergic projections from the basal forebrain to frontal, parietal, temporal, occipital and cingulate cortices: a combined fluorescent tracer and acetylcholinesterase analysis. Brain Res Rev 8:727–749.

Birkmayer W, Hornykiewicz O. 1961. Der L-3, 4-Dioxyphenyl-alanin (DOPOA) Effekt bei der Parkinson-Akiniese. Wien Klin Wschr 73:787–788.

Blessing WW. 1997. Inadequate framework for understanding bodily homeostasis. Trend Neurosci 20:235–239.

Bloom FE. 1993. Advancing a neurodevelopmental origin for schizophrenia. Arch Gen Psychiatr 50:224–227.

Bolam JP, Smith Y. 1990. The GABA and substance P input to dopaminergic neurones in the substantia nigra of the rat. Brain Res 529:57–78.

Bonthius DJ, Solodkin A, Van Hoesen GW. 2005. Pathology of the insular cortex in Alzheimer's disease depends on cortical architecture. J Neuropath and Exp Neurol 64:910–922.

Bouchet M, Cazauvieilh M. 1825. De l'épilepsie considérée dans se reapports avec l'aliénation mentale. Archives Général de Medicine 9:510–542; 10:5–50.

Bouilleret V, Dupont S, Spelle L, Baulac M, et al. 2002. Insular cortex involvement in mesiotemporal lobe epilepsy: a PET study. Annals of Neurology 51:202–208.

Boulay D, Duterte-Boucher D, Leroux-Nicollet I, Naudon L, Costentin J. 1996. Loco-motor sensitization and decrease in [^{3}H] mazindol binding to the dopamine trans-porter are delayed after chronic treatments by GBR12783 or cocaine. J Pharmacol Exp Therap 278:330–337.

Braak H. 1976. A primitive gigantopyramidal field buried in the depth of the cingulate sulcus of the human brain. Brain Res 109:219–233.

Braak H, Braak E. 1991. Neuropathological staging of Alzheimer-related changes. Acta Neuropathol 82:239–259.

Breiter HC, Gollub RL, Weisskoff RM, et al. 1997. Acute effects of cocaine on human brain activity and emotion. Neuron 19:591–611.

Broca P. 1878. Anatomie comparée des circonvolutions cérébrales. Le grand lobe limbique et la scissure limbique dans la série des mammiféres. Rev Anthrop 1: 385–498.

Brockhaus H. 1938. Zur normalen und pathologischen Anatomie des Mandelkerngebietes. J Psychol Neurol (Lpz) 49:1–136.

Brodal A. 1947. The amygdaloid nucleus in the rat. J Comp Neurol 87:1–16.

Brodal A. 1957. The Reticular Formation of the Brain Stem. Anatomical Aspects and Functional Correlations. Edinburgh: Oliver and Boyd.

Brodal A. 1969. Neurological anatomy in relation to clinical medicine. New York: Oxford University Press.

Brodmann K. 1909. Vergleichende Loklisationslehre der Grosshirnrinde, Barth, Leipzig.

Brog JS, Salypongse A, Deutch AY, Zahm DS. 1993. The patterns of afferent innervation of the core and shell in the "accumbens" part of the rat ventral striatum: Immunohistochemical detection of retrogradely transported Fluoro-Gold. J Comp Neurol 338:255–278.

Bubser M, Scruggs JL, Young CD, Deutch AY. 2000. The distribution and origin of the calretinin-containing innervation of the nucleus accumbens of the rat. Eur J Neurosci 12:1591–1598.

Bulfone A, Puelles L, Porteus MH, Frohman MA, Martin JR, Rubenstein J. 1993. Spatially restricted expression of Dlx-1, Dlx-2 (Tes-1), Gbx-2, and Wnt-3 in the embryonic day 12.5 mouse forebrain defines potential transverse and longitudinal segmental boundaries. J Neurosc 13:3155–3172.

Burdach KF. 1819. Vom Baue und Leben des Gehirns. Leipzig.

Butler AB, Hodos W. 1996. Comparative vertebrate neuroanatomy: evolution aand adaptation. New York: Wiley-Liss.

Campbell AW. 1905. Histological Studies on the Localisation of Cerebral Function, Cambridge University Press, London.

Cannon WB. 1929. Bodily changes in pain, hunger, fear, and rage. An account of recent researches into the function of emotional excitement. 2nd ed. New York: Appleton.

Canteras NS, Simerly RB, Swanson LW. 1995. Organization of projections from the medial nucleus of the amygdala: a PHAL study in the rat. J Comp Neurol 360: 213–245.

Cardinal RN, Parkinson JA, Hall J, Everitt BJ. 2002. Emotion and motivation: the role of the amygdala, ventral striatum and prefrontal cortex. Neurosci Biobehav Rev 26:321–352.

Carlsen J, Záborszky L, Heimer L. 1985. Cholinergic projections from the basal forebrain to the basolateral amygdaloid complex: a combined retrograde fluorescent and immunohistochemical study. J Comp Neurol 234:155–167.

Carlsen J, Heimer L. 1988. The basolateral amygdaloid complex as a cortical-like structure. Brain Res 441:377–380.

Carlsson A. 2002. Treatment of Parkinson's with L-DOPA. The early discovery phase, and a comment on current problems. J Neurotrans 109:777–787.

Carlsson A, Lindquist M. 1963. Effect of chlorpromazine or haloperidol on formation of 3-methoxytyramine and normetanephrine in mouse brain. Acta Pharmacol 20:140–144.

Carlsson A, Lindqvist M, Magnusson T, Waldeck B. 1958. On the presence of 3-hydroxytyramine in brain. Science 127:471.

Carmichael ST, Clugnet MC, Price JL. 1994. Central olfactory connections in the Macaque monkey. J Comp Neurol 346:403–434.

Cassell MD. 1998. The amygdala: myth or monolith? TINS 21:200–201.

Cassell MD, Freedman LJ, Shi C. 1999. The intrinsic organization of the central extended amygdala. Ann NY Acad Sci 877:217–242.

Cavanna AE, Trimble MR. 2006. The precuneus: a review of its functional anatomy and behavioral correlates. Brain 129:564–583.

Cereda C, Ghilka J, Maeder P, Bogousslavsky J. 2002. Strokes restricted to the insular cortex. Neurology 59:1950–1955.

Charney DS, Nestler EJ, Bunney BS. 1999. Neurobiology of mental illness. New York: Oxford University Press.

Chikama M, McFarland NR, Amaral DG, Haber SN. 1997. Insular cortical projections to functional regions of the striatum correlate with cortical cytoarchitecture organization in the primate. J Neurosci 17:9686–9705.

Chronister RB, DeFrance JF. 1980. The Neurobiolgy of the Nucleus Accumbens. Brunswick: Haer Institute for Electrophysiological Research.

Cohen DH, Duff TA, Ebbesson SO. 1973. Electrophysiological identification of a visual area in shark telencephalon. Science 182:492.

Cory GA. 2002. Reappraising MacLean's triune brain concept. In: Cory GA, Gardner R, editors. The evolutionary neuroethology of Paul MacLean. Westport: Praeger, pp. 9–28.

Crespo-Facorro B, Kim J, Andreasen NC, O'Leary DS, et al. 2000. Insular cortex abnormalities in schizophrenia. Schizophrenia Research 46:35–43.

Cummings J. Subcortical Dementia. 1990. Oxford University Press, Oxford.

Cummings J, Duchen LW. 1981. Klüver-Bucy syndrome in Pick's Disease, clinical and pathological correlations. Neurology 31:1415–1422.

Cummings J, Trimble MR. 1981. Neuropsychiatric disturbances following brainstem lesions. British Journal of Psychiatry 138:56–59.

Dahlström A, Fuxe K. 1964. Evidence for the existence of monoamine-containing neurons in the central nervous system. I. Demonstration of monoamines in the cell bodies of brain stem neurones. Acta Physiol Scan 62 (Suppl. 232):1–55.

Damasio AR. 1994. Descarte's Error: Emotion, reason and the human brain. Grosset/Putnam Book, New York.

Damasio AR. 1996. The somatic marker hypothesis and the possible functions of the prefrontal cortex. Proc of the Royal Society of London 351:1413–1420.

Damasio AR. 1998. Emotion in the perspective of an integrated nervous system. Brain Res Rev 26:83–86.

Damasio AR, Grabowski T, Bechara A, Damasio H, Ponto LLB, Parvizi J, Hichwa RD. 2000. Subcortical and cortical brain activity during the feeling of self-generated emotions. Nature Neuroscience 3(10):1049–1056.

Damasio H, Grabowski T, Frank R, Galaburda AM, Damasio AR. 1994. The return of Phineas Gage: clues about the brain from the skull of a famous patient. Science 264:1102–1105.

Davey GC, Levy S. 1998. Catstrophic worrying: personal inadequacy and a perseverative interative style as features of the catastrophizing process. J Abnorm Psychol 107:576–586.

Davis M, Falls WA, Campeau S, Kim M. 1993. Fear-potentiated startle: a neural and pharmacological analysis. Behav Brain Res 58:175–198.

Davis M, Shi C. 1999. The extended amygdala: Are the central nucleus of the amygdala and the bed nucleus of the stria terminalis differentially involved in fear and anxiety? Ann N Y Acad Sci 877:292–308.

Davis M, Whalen PJ. 2001. The amygdala: vigilance and emotion. Mol Psychiatr 6:13–34.

de Olmos JS. 1969. A cupric-silver method for impregnation of terminal axon degeneration and its further use in staining granular argyrophilic neurons. Brain Behav Evol 2:213–237.

de Olmos JS. 1972. The amygdaloid projection field in the rat as studied with the cupric silver method. In: Eleftheriou BE, editor. The Neurobiology of the Amygdala. New York: Plenum Press, pp. 145–204.

de Olmos JS. 1990. Amygdala In: Paxinos G, editor. The Human Nervous System. Sydney: Academic Press, pp. 583–710.

de Olmos JS. 2004. Amygdala In: Paxinos G, editor. The Human Nervous System. Amsterdam: Elsevier Academic Press, pp. 739–868.

de Olmos JS, Alheid GF, Beltramino CA. 1985. Amygdala. In: Paxinos G, editor. The Rat Nervous System. Sydney: Academic Press, pp. 223–334.

de Olmos J, Heimer L. 1999. The concepts of the ventral striatopallidal system and extended amygdala. In: McGinty JF, editor. Advancing from the Ventral Striatum to the Extended Amygdala: Implications for Neuropsychiatryand Drug Abuse. New York: New York Academy of Sciences, pp. 1–31.

Degroot A, Parent MB. 2001. Infusions of physostigmine into the hippocampus or the entorhinal cortex attenuate avoidance retention deficits produced by intra-septal infusions of the GABA agonist muscimol. Brain Res 920:10–18.

Degroot A, Kashluba S, Treit D. 2001. Septal GABAergic and hippocampal cholinergic systems modulate anxiety in the plus-maze and shock-probe tests. Pharmacol Biochem Behav 69:391–399.

DelBello MP, Strakowski MD, Zimmerman ME, Hawkins JM, Sax KW. 1999. MRI analysis of the crebellum in bipolar disorder: a pilot study. Neuropsychopharmacology 21(1):63–68.

Dempsey EW, Rioch DM. 1939. The localization in the brain stem of the oestrus responses of the female guinea pig. J Neurophysiol 138:283–296.

Devinsky O, Morrell M, Vogt BA. 1995. Contributions of the anterior cingulate cortex to behaviour. Brain 118:279–306.

DiCara LV. 1974. Limbic and Autonomic Nervous Systems Research. New York: Plenum.

Ding S-L, Morecraft RJ, Van Hoesen GW. 2003. The topography, cytoarchitecture and cellular phenotypes of cortical areas that form the cingulo-parahippocampal isthmus and adjoining retrocalcarine areas in the monkey. J Comp Neurol 456:184–201.

Doane BK, Livingston KE. 1986. The Limbic System: Functional Organization and Clinical Disorders. New York: Raven.

Dong HW, Petrovich GD, Swanson LW. 2001. Topography of projections from amygdala to the bed nuclei of the stria terminalis. Brain Res Rev 38:192–246.

Dong H-W, Petrovich GD, Watts AG, Swanson LW. 2001. Basic organization of projections from the oval and fusiform nuclei of the bed nuclei of the stria terminalis in adult rat brain. J Comp Neurol 436:430–455.

Dresel K. 1924. Klin Wochenschr 2:2231.

Drevets W. 2003. Neuroimaging abnormalities in the amygdala in mood disorders. In: The Amygdala in Brain Function, Basic and Clinical Approaches. Ed Shinnick-Gallagher P, et al. Annals of the New York Academy of Sciences, Vol 985, New York, pp. 420–444.

Drevets WC, Price JL, Simpson JR, et al. 1997. Subgenual prefrontal cortex abnormalities in mood disorders. Nature, 386:824–827.

Drugan RC, Skolnick P, Paul SM, Crawley JN. 1986. Low doses of muscimol produce anticonflict actions in the lateral septum of the rat. Neuropharmacology 25:203–205.

Duman RS, Malberg J, Nakagawa S, D'Sa C. 2000. Neuronal plasticity and survival in mood disorders. Biol Psychiatry 48:732–739.

Durant JR. 1985. The science of sentiment: the problem of the cerebral localization of emotion. In: Bateson PPG, Klopfer PH, editors. Perspectives in ethology. New York, Plenum Press, pp. 1–31.

Ebbesson SO, Heimer L. 1970. Projections of the olfactory tract fibers in the nurse shark (Ginglymostoma cirratum). Brain Res 17:47–55.

Ebbesson SO, Schroeder DM. 1971. Connections of the nurse shark's telencephalon. Science 173:254.

Ebbesson SO. 1980. Comparative neurology of the telencephalon. New York: Plenum.

Ehringer H, Hornykiewicz O. 1960. Verteilung von Noradrenalin and Dopamin (3-Hydroxytryptamin) im Gehirn des Menschen un ihr Verhalten be Erkrankungen des extrapyramidalen Systems. Klin Wochen shcr 38:1236–1239.

Eriksson PS, Perfilieva E, Bjork-Eriksson T, Alborn A-M, Nordborg C, Peterson DA, Gage FH. 1998. Neurogenesis in the adult hippocampus. Nature Medicine 4:1313–1317.

Eslinger PJ, Grattan LJ. 1993. Frontal lobe and frontal-striatal substrates for different forms of human cognitive flexibility. Neuropsychologia. 31:17–28.

Falck B, Hillarp N-Å, Thieme G, Torp A. 1962. Fluorescence of catecholeamines and related compounds condenced with formaldehyde. J Histochem Cytochem 10:348–354.

Fallon JH, Loughlin SE. 1985. Monoamine innervation of cerebral cortex and a theory of the role of monoamines in cerebral cortex and basal ganglia. Cerebral Cortex 6:41–127.

Ferrier D. 1876/1966. *The Functions of the Brain*. Smith Elder, London, 1976; reprinted in 1966 by Dawsons of Pall Mall, London.

Ferry AT, Ongur D, An X, Price JL. 2000. Prefrontal cortical projections to the striatum in macaque monkeys. Evidence for an organization related to prefrontal networks. J Comp Neurol 425:447–470.

Filimonoff IN. 1947. A rational subdivision of the cerebral cortex. Arch Neuro Psychiatr 58:296–311.

Fink RP, Heimer L. 1967. Two methods for selective silver impregnation of degenerating axons and their synaptic endings in the central nervous system. Brain Research 4:369–374.

Fox CA. 1943. The stria terminalis, longitudinal assocciation bundle and precommissural fornix in the cat. J Comp Neurol 79:277–295.

Frankel AH. 1978. Sequential response shift rate: a correlate of human adaptivity measureable with existing personality inventories. J Psychol 98 (1st half):129–143.

Frodl T, Schaub A, Banac, et al. 2006. Reduced hippocampal volume correlates with executive dysfunctioning in major depression. Journal of Psychiatry and Neuroscience 31:316–323.

Frodl T, Schaub A, Banac S, et al. 2006. Reduced hippocampal volume correlates with executive dysfunctioning in major depression. Journal of Psychiatry and Neuroscience 31:316–323.

Fudge JL, Kunishio K, Walsh P, Richard C, Haber SN. 2002. Amygdaloid projections to ventro-medial striatal subterritories in the primate. Neuroscience 110:257–275.

Fuster JM. 2003. Cortex and Mind. Oxford University Press, Oxford.

Gallagher M, Graham PW, Holland PC. 1990. The amygdala central nucleus and appetitive Pavlovian conditioning: lesions impair one class of conditioned behavior. J Neurosci 10:1906–1911.

Gallagher M, Holland PW. 1994. The amygdala complex: multiple roles in associative learning and attention. PNAS 91:11771–11776.

Gardner R. 2002. MacLean's paradigm and its relevance for psychiatry's basic science. In: Cory GA, Gardner R, editors. The evolutionary neuroethology of Paul MacLean. Westport: Praeger, pp. 85–105.

Gastard MC, Jensen SL, Martin JR III, Williams EA, Zahm DS. 2002. The caudal sublenticular region/anterior amygdaloid area is the only part of the rat forebrain and mesopontine tegmentum occupied by magnocellular cholinergic neurons that receives outputs from the central division of extended amygdala. Brain Res 957:207–222.

Gazzaniga MS. 2004. The cognitive neurosciences. Cambridge: MIT Press.

Geisler S, Andres KHW, Veh RW. 2003. Cytochemical criteria for the identification of subnuclei in the rat lateral habenular complex. J Comp Neurol 458:78–97.

Geisler S, Derst C, Veh RW, Zahm DS. 2007. Glutamatergic afferents of the ventral tegmental area in the rat. J Neurosci 27:5730–5743.

Geisler S, Zahm DS. 2005. Afferents of the ventral tegmental area in the rat—anatomical substratum for integrative functions. J Comp Neurol 490:270–294.

George M, Ketter TA, Parekh PL, et al. 1997. Blunted left cingulate activation in mood disorder subjects during a response interference task. Journal of Neuropsychiatry and Clinical Neuroscience 9:55–63.

Gerfen CR. 1992. The neostriatal mosaic: multiple levels of compartmental organization. Trends Neurosci 15:133–139.

Gloor P. 1997. The Temporal Lobe and Limbic System. Oxford University Press, New York.

Goldby F. 1937. An experimental investigation of the cerebral hemispheres of Lacerta viridis. J Anat 71:332–355.

Golgi C. 1873. Sulla struttura della grigia del cervello. Gazetta Medical Italiani Lombardi 6:244–246. Translated by M. Santini as "On the structure of the gray matter of the brain," in M. Santini (Ed.), *Golgi Centennial Symposium, Proceedings*. New York: Raven Press, 1975. pp. 647–650.

Gould E, Reeves AJ, Graziano MSA, Gross CG. 2006. Neurogenesis in the neocortex of adult primates. Science 286:548–552.

Gray JA. 1982. The Neuropsychology of Anxiety. Oxford University Press, Oxford.

Graybiel AM, Ragsdale CW. 1983. Biochemical anatomy of the striatum. In Emson PC, editor. Chemical Neuroanatomy. New York: Raven, pp. 427–504.

Grimm JW, Chapman MA, Zahm DS, See RE. 2001. Decreased choline acetyltransferase immunoreactivity in discrete striatal subregions following chronic haloperidol in rats. Synapse 39:51–57.

Gritti I, Mainville L, Jones BE. 1993. Codistribution of GABA-with acetylcholine-synthesizing neurons in the basal forebrain of the rat. J Comp Neurol 329:438–457.

Gritti I, Mainville L, Mancia M, Jones BE. 1997. GABAergic and other noncholinergic basal forebrain neurons, together with cholinergic neurons, project to the mesocortex and isocortex in the rat. J Comp Neurol 383:163–177.

Groenewegen HJ, Berendse HW. 1994. Anatomical relationships between the prefrontal cortex and the basal ganglia in the rat. In: Thierry AM, Glowinski J, Goldman-Rakic PS, Christen Y, editors. Motor and Cognitive Functions of the Prefrontal Cortex. Berlin, Heidelberg: Springer-Verlag, pp. 31–77.

Groenewegen HJ, Berendse HW, Haber SN. 1993. Organization of the output of the ventral striatopallidal system in the rat: ventral pallidal efferents. Neurosci 57:113–142.

Groenewegen HJ, Berendse HW, Wolters JG, Lohman AH. 1990. The anatomical relationship of the prefrontal cortex with the striatopallidal system, the thalamus and the amygdala: evidence for a parallel organization. Prog Brain Res 85:95–116; discussion 116–118.

Groenewegen HJ, Berendse HW, Wouterlood FG. 1994. Organization of the projections from the ventral striato-pallidal system to ventral mesencephalic dopaminergic neurons in the rat. In: Percheron G, McKenzie JS, Féger J, editors. The Basal Ganglia IV. New York: Plenum Press, pp. 81–93.

Groenewegen HJ, Russchen FT. 1984. Organization of the efferent projections of the nucleus accumbens to pallidal, hypothalamic, and mesencephalic structures: a tracing and immunohistochemical study in the cat. J Comp Neurol 223:347–367.

Groenewegen HJ, Wright CI, Uylings HB. 1997. The anatomical relationships of the prefrontal cortex with limbic structures and the basal ganglia. J Psychopharmacol 11:99–106.

Grove EA. 1988. Efferent connections of the substantia innominata in the rat. J Comp Neurol 277:347–364.

Grove EA, Domesick VB, Nauta WJH. 1986. Light microscopic evidence of striatal input to intrapallidal neurons of cholinegic cell group Ch4 in the rat: a study employing the anterograde tracer Phaseolus vulgaris leucoagglutinin (PHA-L). Brain Res 367:379–384.

Gurdjian EA. 1928. The corpus striatum of the rat. Studies on the brain of the rat, No. 3. J Comp Neurol 45:249–281.

Gurevich EV, Bordelon Y, Shapiro RM, Arnold SE, Gur RE, Joyce JN. 1997. Meso-limbic dopamine D3 receptors and use of antipsychotics in patients with schizo-phrenia. A postmortem study. Arch Gen Psychiatry 54:225–232.

Haber SN. 2003. Integrating cognition and motivation into the basal ganglia pathways of action. In: Bédard MA, Agid Y, Chouinard S, Fahn S, Korczyn AD, Lespérance P, editors. Mental and behavioral dysfunction in movement disorders. Totowa: Humana Press, pp. 35–50.

Haber SN, Fudge JL, McFarland NR. 2000. Striatonigrostriatal pathways in primates form an ascending spiral from the shell to the dorsolateral striatum. J Neurosc 20:2369–2382.

Haber SN, Johnson Gdowsky M. 2004. The basal ganglia. In: Paxinos G, Mai JK, editors. The human nervous system. Amsterdam: Elsevier. pp. 676–738.

Haber SN, Kunishio K, Mizobuchi M, Lynd-Balta E. 1995. The orbital and medial prefrontal circuit through the primate basal ganglia. J Neurosci 15:4851–4867.

Haber SN, Lynd E, Klein C, Groenewegen HJ. 1990. Topographic organization of the ventral striatal efferent projections in the Rhesus monkey: an anterograde tracing study. J Comp Neurol 293:282–298.

Haberly LB, Price JL. 1978. Association and commissural fiber systems of the olfactory cortex of the rat. J Comp Neurol 178:711–740.

Hallanger AE, Wainer BH. 1988. Ascending projections from the pedunculopontine tegmental nucleus and the adjacent mesopontine tegmentum in the rat. J Comp Neurol 274:483–515.

Han J-S, McMahan RW, Holland P, Gallagher M. 1997. The role of an amygdalo-nigrostriatal pathway in associative learning. J Neuroscience 17:3913–3919.

Harrington A. 1991. At the intersection of knowledge and values: fragments of a dialogue in Woods Hole, Massachusetts, August 1990. In: Harrington A, editor. So human a brain. Boston: Birkhäuser; pp. 247–324.

Harris GW. 1958. Chairman's opening remarks. In: Wolstenholme GEW, O'Connor CM, eds. Ciba Foundation Symposium on the Neurological Basis of Behavior. J & A Churchill, LTD, London, pp. 1–3.

Hasue RH, Shammah-Lagnado SJ. 2002. Origin of the dopaminergic innervation of the central extended amygdala and accumbens shell: a combined retrograde tracing and immunohistochemical study in the rat. J Comp Neurol, 454:15–33.

Hawkes C. 2003. Olfaction in neurodegenerative disorder. Movement Disorders, 18: 364–372.

Hayakawa T, Zyo K. 1983. Comparative cytoarchitectonic study of Gudden's temental nuclei in some mammals. J Comp Neurol 216:233–244.

Heath RG. 1954. Studies in schizophrenia. Cambridge, Massachusetts: Harvard University Press.

Heath RG, Cox AW, Lustick LS. 1974. Brain activity during emotional states. Am J Psychiatry 131:858–862.

Heimer L. 1969. The secondary olfactory connctions in mammals, reptiles and sharks. Ann N Y Acad Sci 167:129–146.

Heimer L. 1970. Bridging the gap between light and electron microscopy in the experimental tracing of fiber connections. In: Nauta WHJ and Ebbesson SOE, editors. Contemporary research methods in neuroanatomy. New York: Springer-Verlag. pp. 162–172.

Heimer L. 1972. The olfactory connections of the diencephalon in the rat. An experimental light- and electron-microscopic study with special emphasis on the problem of terminal degeneration. Brain Behav Evol 6:484–523.

Heimer L. 2000. Basal forebrain in the context of schizophrenia. Brain Res Rev 31:205–235.

Heimer L. 2003. The legacy of the silver methods and the new anatomy of the basal forebrain: implications for neuropsychiatry and drug abuse. Scand J Psychol 44:189–201.

Heimer L. 2003. A new anatomical framework for neuropsychiatric disorders and drug abuse. Am J Psychiat 160:1726–1739.

Heimer L, Alheid GF. 1991. Piecing together the puzzle of basal forebrain anatomy. In: Napier TC, Kalivas PW, Hanin I, editors. The Basal Forebrain: Anatomy to Function. New York: Plenum Press, pp. 1–42.

Heimer L, Alheid GF, de Olmos JS, Groenewegen HJ, Haber SN, Harlan RE, Zahm DS. 1997. The accumbens: beyond the core-shell dichotomy. J Neuropsychiat Clin Neurosci 9:354–381.

Heimer L, Alheid GF, Zahm DS. 1993. Basal forebrain organization: An anatomical framework for motor aspects of drive and motivation. In: Kalivas PW, editor. The Mesolimbic Motor Circuit and Its Role in Neuropsychiatric Disorders. Boca Raton, FL: CRC Press; pp. 1–44.

Heimer L, de Olmos JS, Alheid GF, Záborszky L. 1991. "Perestroika" in the basal forebrain: opening the border between neurology and psychiatry. Prog Brain Res 87:109–165.

Heimer L, de Olmos J, Alheid GF, Pearson J, Sakamoto M, Marksteiner J, Switzer III RC. 1999. The human basal forebrain, part 2. In Bloom FE, Bjorklund A, Hokfelt T, editors. Handbook of Chemical Neuroanatomy, Vol. 15. Amsterdam: Elsevier, pp. 57–226.

Heimer L, Harlan RE, Alheid GF, Garcia M, de Olmos J. 1997. Substantia innominata: a notion which impedes clinical-anatomical correlations in neuropsychiatric disorders. Neuroscience 76:957–1006.

Heimer L, RoBards MJ (eds.). 1981. Neuroanatomical Tract-tracing Methods, Plenum Press, New York and London.

Heimer L, Switzer RC, Van Hoesen GW. 1982. Ventral striatum and ventral pallidum: additional components of the motor system? Trends Neurosci 5:83–87.

Heimer L, Wilson RD. 1975. The subcortical projections of allocortex: Similarities in the neuronal associations of the hippocampus, the piriform cortex and the neocortex. In: Santini M, editor. Golgi Centennial Symposium Proceedings. New York: Raven Press, pp. 173–193.

Heimer L, Van Hoesen GW. 2006. The limbic lobe and its output channels: Implications for emotional functions and adaptive behavior. Neurosci Biobehav Rev 30:126–147.

Heimer L, Van Hoesen GW, Rosene DL. 1977. The olfactory pathways and the anterior perforated substance in the primate brain. Internatl J Neurol 12:42–52.

Heimer L, Záborszky L (eds.). 1989. Neuroanatomical Tract-tracing Methods 2, Plenum Press, New York and London.

Heimer L, Záborszky L, Zahm DS, Alheid GF. 1987. The ventral striatopallidothalamic projection: I. The striatopallidal link originating in the striatal parts of the olfactory tubercle. J Comp Neurol 255:571–591.

Heimer L, Zahm DS, Alheid GF. 1995. Basal ganglia. In: Paxinos G, editor. The Rat Nervous System, 2nd edition. San Diego, CA: Academic Press, pp. 579–628.

Heimer L, Zahm DS, Churchill L, Kalivas PW, Wohltmann C. 1991. Specificity in the projection patterns of accumbal core and shell in the rat. Neuroscience 41:89–125.

Hendersen Z. 1997. The projection from the striatum to the nucleus basalis in the rat: an electron microscopic study. Neurosci 78:943–955.

Herbert J. 1997. Do we need a limbic system? Trend Neurosci 20:508–509.

Herkenham M. 1979. The afferent and efferent connections of the ventromedial thalamic nucleus in the rat. J Comp Neurol 183:487–518.

Herkenham M. 1986. New perspectives on the organization and evolution of nonspecific thalamocortical projections. In: Jones EG, Peters A, editors. Cerebral Cortex. New York: Plenum Press, pp. 403–445.

Herkenham M, Nauta WJH. 1977. Afferent connections of the habenular nuclei in the rat. A horseradish peroxidase study, with a note on the fiber-of-passage problem. J Comp Neurol 173:123–146.

Herkenham M, Nauta WJH. 1979. Efferent connections of the habenular nuclei in the rat. J Comp Neurol 187:19–48.

Herrick CJ. 1926. Brains of rats and men. Chicago: University of Chicago Press.

Herrick CJ. 1956. The evolution of human nature. University of Texas, Austin.

Herzog AG, Van Hoesen GW. 1976. Temporal neocortical afferent connections to the amygdala in the rhesus monkey. Brain Res 115:57–69.

Holstege G. 1989. Descending motor pathways and the spinal motor system: Limbic and non-limbic components. Prog Brain Res 87:307–421.

Holstege G. 1992. The emotional motor system. Eur J Neurosci 30:67–79.

Holstege G, Bandler R, Saper CB. 1996. The Emotional Motor System. Progress in Brain Research, Volume 107. Amsterdam: Elsevier.

Hotz G, Helm-Estabrooks N. 1995. Perseveration. Part I: a review. Brain Inj 9:151–159.

Hyman BT, Van Hoesen GW, Damasio AR, Barnes CL. 1984. Alzheimer's disease: Cell specific pathology isolates the hippocampal formation. Science 225:1168–1170.

Inglis WL, Winn P. 1995. The pedunculopontine tegmental nucleus: where the striatum meets the reticular formation. Progress in Neurobiol 47:1–29.

Insausti R, Amaral DG, Cowan WM. 1987. The entorhinal cortex of the monkey, II: Cortical afferents. J Comp Neurol 264:326–355.

Insausti R, Tunon T, Sobreviela T, Insausti AM, Gonzalo LM. 1995. The human entorhinal cortex: a cytoarchitectonic analysis. J Comp Neurol 355:171–198.

Isaacson RL. 1982. The Limbic System. 2nd ed. New York: Plenum.

Isaacson RL. 1992. A fuzzy limbic system. Behav Brain Res 52:129–131.

Jakob H, Beckman H. 1986. Prenatal developmental disturbances in the limbic allocortex of schizophrenics. J Neural Transm 65:303–326.

Jakab RL, Leranth C. Septum. In: Paxinos G, editor. The Rat Nervous System, Volume 1, Forebrain and Midbrain, 2nd Edition, Sydney: Academic Press; 1995, pp. 405–442.

James W. 1884. What is an emotion? Mind 9:188–205.

James W, Lange CG. 1922. The emotions. Baltimore: Williams and Wilkins.

Jeste DV, Lohr JB. 1989. Hippocampal pathologic findings in schizophrenia. A morphometric study. Arch Gen Psychiatry, 46:1019–1024.

Joel D, Weiner I. 1994. The organization of the basal ganglia-thalamocortical circuits: Open interconnected rather than closed segregated. Neuroscience 63:363–379.

Johnston JB. 1923. Further contribution to the study of the evolution of the forebrain. J Comp Neurol 35:337–481.

Johnston M, McKinney M, Coyle JT. 1979. Evidence from a cholinergic projection to neocortex from neurons in the basal forebrain. Proc Natl Acad Sci USA 76: 5392–5396.

Jones BE. 1995. Reticular formation. In: Paxinos G, editor. The Rat Nervous System. Second Edition. New York: Academic Press; pp. 155–171.

Jones DL, Mogenson GJ. 1980. Nucleus accumbens to globus pallidus GABA projection subserving ambulatoryy activity. Am J Physiol 238:65–69.

Jones EG, Burton H, Saper CB, Swanson LW. 1976. Midbrain, diencephalic and cortical relationships of the basal nucleus of Meynert and associated structures in primates. J Comp Neurol 167:385–420.

Jones EG, Powell TP. 1970. An anatomical study of converging sensory pathways within the cerebral cortex of the monkey. Brain 93:793–820.

Joseph R. 1999. Frontal lobe psychopathology: mania, depression, confabulation catatonia, perseveration, obsessive compulsions, and schizophrenia. Psychiatry 62: 138–172.

Joyce JN, Gurevich EV. 1999. D3 receptors and the actions of neuroleptics in the ventral striatopallidal system of schizophrenics. Ann N Y Acad Sci 877:595–613.

Joyce JN, Millan MJ. 2005. Dopamine D 3 receptor anatagonists as therapeutic agents. www.drugdiscoverytoday.com 10:917–925.

Jurgens U, Ploog D. 1970. Cerebral representation of representation in the squirrel monkey. Experimental Brain Research 10:532–554.

Kalivas PW, Churchill L, Romanides A. 1999. Involvement of the pallidal-thalamocortical circuit in adaptive behavior. In: McGinty JF, editor. Advancing from the ventral striatum to the extended amygdala. New York: New York Acad Sci pp. 64–70.

Kanemoto K, Kawasaki J, Kawai I. 1996. Post-ictal psychosis: a comparison with acute interictal and chronic psychoses. Epilepsia 37:551–556.

Kapp BS, Pascoe JP, Bixter MA. 1984. The amygdala: A neuroanatomical systems approach to its contribution to aversive conditioning. In: Squire LR, Butters N (Eds.), Neuropsychology of Memory. The Guilford Press, New York, pp. 473–488.

Karten HJ. 1969. The organization of the avian telencephalon and some speculations on the phylogeny of the amniote telencephalon. Ann N Y Acad Sci 167:164–179.

Keay KA, Bandler R. 2001. Parallel circuits mediating distinct emotional coping reactions to different types of stress. Neurosci Biobehav Rev 25:669–678.

Kelley AE. 2004. Ventral striatal control of appetitive motivation: role in ingestive behavior and reward-related learning. Neurosci Biobehav Rev 27:765–776.

Kelley AE, Smith-Roe SL, Holahan MR. 1997. Response-reinforcement learning is dependent on N-methyl-D-aspartate receptor activation in the nucleus accumbens core. Proc Natl Acad Sci USA 94(22):12174–12179.

Kelly PH, Seviour PW, Iversen SD. 1975. Amphetamine and apomorphine responses in the rat following 6-OHDA lesions of the nucleus accumbens septi and corpus striatum. Brain Res 94:507–522.

Kelly PH, Roberts CS. 1983. Effects of amphetamine and apomorphine on locomotor activity after 6-OHDA and electrolytic lesions of the nucleus accumbens. Pharm Biochem Behav 19:137–143.

Kemp JM, Powell TPS. 1971. The connexions of the striatum and globus pallidus: synthesis and speculation. Phil Trans R Soc Lond B 262:441–457.

Kievet J, Kuypers HGJM. 1977. Organization of thalamo-cortical connexions to the frontal lobe in the rhesus monkey. Exp Brain Res 29:299–322.

Kita H, Kitai ST. 1987. Efferent projections of the subthalamic nucleus in the rat: light and electron microscopic analysis with the PHA-L method. J Comp Neurol 260:435–452.

Kitayama N, Vaccarino V, Kutner M, Weiss P, Bremner JD. 2005. Magnetic resonance imaging of hippocampal volume in PTSD: a metaanalysis. Journal of Affective Disorders 88:79–86.

Klüver H, Bucy P. 1937. "Psychic blindness" and other symptoms following bilateral temporal lobectomy in rhesus monkey. Am J Physiol 119:352–353.

Koella WP, Trimble MR. 1982. Temporal Lobe Epilepsy, Mania, and Schizophrenia and the Limbic System. Basel: Karger.

Koestler A. 1967. The ghost in the machine. London: Hutchinson.

Koob GF. 1999. The role of the striatopallidal and extended amygdala systems in drug addiction. In: McGinty JF, editor. Advancing from the ventral striatum to the extended amygdala. New York: New York Acad Sci pp. 445–460.

Koob GF, Le Moal M. 2006. Neurobiology of addiction. Amsterdam: Elsevier.

Koob GF, Stinus L, Le Moal M. 1981. Hyperactivity and hypoactivity produced by lesions to the mesolimbic dopamine system. Behav Brain Res 3:341–359.

Koos T, Tepper JM. 1999. Inhibitory control of neostriatal projections neurons by GABAergic interneurons. Nature Neuroscience 2:467–472.

Kopelman A, Andreasen N, Nopoulos P. 2005. Morphology of the anterior cingulate gyrus in patients with schizophrenia: relationship to typical neuroleptic exposure. American Journal of Psychiatry 162:1872–1878.

Kötter R, Meyer N. 1992. The limbic system: a review of its empirical foundatioin. Behav Brain Res 52:105–127.

Krettek JF, Price JL. 1978. Amygdaloid projections to subcortical structures within the forebrain and brainstem in the rat and cat. J Comp Neurol 178:225–254.

Lautin A. 2001. The limbic brain. New York: Kluwer Academic/Plenum Publishers.

Lawrie SM, Nhalley HC, Job DE, Johnstone EC. 2003. Structural and functional abnormalities of the amygdala in schizophrenia. In: The Amygdala in Brain Function, Basic and Clinical Approaches. Shinnick-Gallagher P, et al., editors Annals of the New York Acadamy of Sciences, Vol 985, New York, pp. 445–460.

LeDoux JE. 1991. Emotion and the brain. The Journal of NIH Research 3:48–51.

LeDoux J. 1996. The Emotional Brain. New York: Simon & Schuster.

LeDoux JE, Iwata J, Cicchetti P, Reis DJ. 1988. Different projections of the central amygdaloid nucleus mediate autonomic and behavioral correlates of conditioned fear. J Neurosci 8:2517–2529.

Lehmann J, Nagy JI, Atmadja S, Fibiger HC. 1980. The nucleus basalis magnocellularis: the origin of a cholinergic projection to the neocortex of the rat. Neurosci 5:1161–1174.

Leontovich TA, Zhukova GP. 1963. The specificity of the neuronal structure and topography of the reticular formation in the brain and spinal cord of carnivora. J Comp Neurol 121:347–379.

Leranth C, Frotscher M. 1989. Organization of the septal region in the rat brain: Cholinergic-GABAergic interconnections and the termination of hippocampo-septal fibers. J Comp Neurol 289:304–314.

Lichter DG, Cummings JL. 2001. Frontal-subcortical circuits in psychiatric and neurological disorders. New York: The Guilford Press.

Livingston KE, Hornykiewicz O. 1978. Limbic Mechanisms: The Continuing Evolution of the Limbic System Concept. New York: Plenum.

Loopuijt LD, Williams EA, Zahm DS. Small cholinergic neurons in the caudal sublenticular region/anterior amygdaloid area in the rat. Soc Neurosci Abstr 2003; Prog # 600.11.

Lucassen PJ, Heine VM, Muller MB, et al. 2006. Stress, depression and hippocampal apoptosis. CNS Neurological Disorder Drug Targets 5:531–546.

MacLean PD. 1949. Psychosomatic disease and the "visceral brain." Recent developments bearing on the Papez theory of emotion. Psychosom. Med. 11:338–353.

MacLean PD. 1952. Some psychiatric implications of physiological studies on frontotemporal portions of limbic system (visceral brain). Electroencephalogr Clin Neurophysiol 4:407–418.

MacLean PD. 1955. The limbic system ("visceral brain") in relation to central gray and reticulum of the brain stem. Psychosomatic Medicine 17:355–366.

MacLean PD. 1958. Constrasting functions of limbic and neocortical systems of the brain and their relevance to psychophysiological aspects of medicine. American J Med 25:611–626.

MacLean PD. 1970. The triune brain, emotion, and scientific bias. In: Schmitt FO, editor. The Neurosciences, second study program. New York: Rockefeller University Press.

MacLean PD. 1978. Challenges of the Papez heritage. In: Livingston KE, Hornykiewicz O, editors. Limbic mechanisms: the continuing evolution of the limbic system concept.l. New York: Plenum Press, pp. 1–15.

MacLean PD. 1990. The Triune Brain in Evolution. New York: Plenum.

Mah L, Zarate CA, Singh J, Duan YF, Luckenbaugh DA, Manji HK, Drevets WC. 2007. Regional cerebral glucose metabolic abnormalities in bipolar 2 depression. Biological Psychiatry 61:765–775.

Manns ID, Mainville L, Jones BE. 2001. Evidence for glutamate, in addition to acetylcholine and GABA, neurotransmitter synthesis in basal forebrain neurons projecting to entorhinal cortex. Neurosci 107:249–263.

Matsumoto M, Hikosaka O. 2007. Lateral habenula as a source of negative reward signals in dopamine neurons. Nature 23 May [Epub ahead of print].

Malamud DN. 1975. Organic Brain Disease mistaken for psychiatric disorder. In: Benson DF, Blumer D, eds. Psychiatric Aspects of Neurological Disease, Vol 2. New York, Grune and Stratton, 287–305.

Martin LJ, Hadfield MG, Dellovade TL, Price DL. 1991. The striatal mosaic in primates: patterns of neuropeptide immunoreactivity differentiate the ventral striatum from the dorsal striatum. Neuroscience 43:397–417.

Matthysse S. 1973. Antipsychotic drug actions: a clue to the neuropathology of schizophrenia? Fed Proc 32:200–205.

Mayberg HS, Lozano AM, Voon V, McNeely HE. 2005. Seminowicz D, Hamani C, Schwalb JM, Kennedy SH. Deep brain stimulation for treatment-resistant depression. Neuron 45:651–660.

Mayberg HS, Liotti M, Brannan SK, McGinnis S, Mahurin RK, Jerabek PA, Silva JA, Tekell JL, Martin CC, Fox PT. 1999. Reciprocal limbic-cortical function and negative mood: converging PET findings in depression and normal sadness. Am J Psychiatry 156:675–682.

McCarthy G, Blamire AM, Rothman DL, Gruetter R, Schulman RG. 1993. Echoplanar MRI studies of frontal cortex activation during word generation in humans. Proceedings of the National Academy of Sciences, 90:4952–4956.

McDonald AJ. 1982. Cytoarchitecture of the central amygdaloid nucleus in the rat. J Comp Neurol 208:401–418.

McDonald AJ. 2003. Is there an amygdala and how far does it extend? In: Shinnick-Gallagher P, Pitkänen A, Shekhar A, Cahill L, editors. The amygdala in brain function. New York: New York Academy of Sciences pp. 1–21.

McDonald AJ, Mascagni F, Guo L. 1996. Projections of the medial and lateral prefrontal cortices to the amygdala: a *Phaseolus vulgaris* leucoagglutinin study in the rat. Neuroscience 71:55–75.

McDonald AJ, Shammah-Lagnado SJ, Shi C, Davis M. 1999. Cortical afferents to the extended amygdala. Ann N Y Acad Sci 877:309–338.

McFarland K, Davidge SB, Lapish CC, Kalivas PW. 2004. Limbic and motor circuitry underlying footshock-induced reinstatement of cocaine-seeking behavior. J Neurosci. 24:1551–1560.

McFarland K, Lapish CL, Kalivas PW. 2003. Prefrontal glutamate release into the core of the nucleus accumbens mediates cocaine reinstatment of drug-seeking behavior. J Neurosci 23:3531–3537.

McGeorge AJ, Faull RLM. 1989. The organization of the projection from the cerebral cortex to the striatum in the rat. Neuroscience 29:503–537.

McGinty JF. 1999. Advancing from the ventral striatum to the extended amygdala. New York: New York Academy of Sciences.

McKinney M, Coyle JT, Hedreen JC. 1983. Topographic analysis of the innervation of the rat neocortex and hippocampus by the basal forebrain cholinergic system. J Comp Neurol 217:103–121.

McIntyre DC, Kelly ME, Staines WA. 1996. Efferent projections of the anterior perirhinal cortex in the rat. J Comp Neurol 369:302–318.

McMullen NT, Almli CR. 1981. Cell types within the medial forebrain bundle: a Golgi study of preoptic and hypothalamic neurons in the rat. Am J Anat 161:323–340.

Meesen H, Olszewski J. 1949. A cytoarchitectonic atlas of the rhombencephalon in the rabbit. Karger, Basel.

Mega MS, Cummings JL. 1994. Frontal-subcortical circuits and neuropsychiatric disorders. J Neuropsychiatry 6:358–370.

Mega MS, Cummings J. 1997. The cingulate and cingulate syndromes. In: Trimble MR, Cummings JL, Contemporary behavioural neurology, Butterworths, Oxford, pp. 189–214.

Meredith GE, Pattiselanno A, Groenewegen HJ, Haber SN. 1996. Shell and core in monkey and human nucleus accumbens identified with antibodies to calbindin-D28k. J Comp Neurol 365:628–639.

Mesulam M-M. 1995. Cholinergic pathways and the ascending reticular activating system of the human brain. Ann NY Acad Sci 757:169–179.

Mesulam M-M. 2000. Neuroplasticity failure in Alzheimer's disease: bridging the gap bestween plaques and tangles. Neuron 24:521–529.

Mesulam M-M. 2000. Principles of Behavioral and Cognitive Neurology. Oxford University Press, Oxford.

Mesulam M-M, Mufson EJ. 1984. Neural inputs into the nucleus basalis of the substantia innominata (Ch4) in the rhesus monkey. Brain 107:253–274.

Mesulam M-M, Mufson EJ. 1985. The insula of Reil in man and monkey. In: Peters A, Jones EG, editors, Cortex. Plenum Press, New York, pp. 179–226.

Mesulam M-M, Mufson EJ, Levey AI, Wainer BH. 1983. The cholinergic innervation of the cortex by the basal forebrain: cytochemistry and cortical connections of the septal area, diagonal band nuclei, nucleus basalis (substantia innominata) and hypothalamus in the rhesus monkey. J Comp Neurol 214:170–197.

Mesulam M-M, Mufson EJ, Wainer BH, Levey AI. 1983. Central cholinergic pathways in the rat: an overview based on an alternative nomenclature (Ch1–Ch6). Neurosci 10:1185–1201.

Mesulam M-M, Van Hoesen GW. 1976. Acetylcholinesterase-rich projections from the basal forebrain of the rhesus monkey to neocortex. Brain Res 109:152–157.

Meyer M, Allison AC. 1949. An experimental investigation of the connexions of the olfactory tracts in the monkey. J Neurol Neurosurg Psychiatr 12:274–286.

Meyer G, Gonzalez-Hernandez T, Carrillo-Padilia F, Ferres-Torres R. 1989. Aggregations of granule cells in the basal forebrain (islands of Calleja). A Golgi and cytoarchitectonic study in different mammals including man. J Comp Neurol 284: 405–478.

Meynert T. 1872. Vom Gehirne der Saugethiere. In: Stricker S, editor. Handbuch der Lehre von den Geweben des Menschen und Thiere. (Transl. into English in A Manual of Histology, by S. Stricker, New York, Wood, 1872.) Leipzig: Engelmann. pp. 694–808.

Middleton FA, Strick PL. 2001. A revised neuroanatomy of frontal-subcortical circuits. In: Lichter DG, Cummings J, editors. Frontal-subcortical circuits in psychiatric and neurological disorders. New York: The Guilford Press. pp. 44–58.

Millhouse OE. 1986. The intercalated cells of the amygdala. J Comp Neurol 247:246–271.

Millhouse O, Heimer L. 1984. Cell configurations in the olfactory tubercle of the rat. J Comp Neurol 228:571–597.

Mitani A, Ito K, Hallanger AE, Wainer BH, Kataoka K, McCarley RW. 1988. Cholinergic projections from the laterodorsal and pedunculopontine tegmental nuclei to the pontine gigantocellular tegmental field in the rat. Brain Res 451:397–402.

Mogenson J, Jones DL, Yim SY. 1980. From motivation to action: functional interface between the limbic system and the motor system. Progress in Neurobiology 14:69–97.

Morecraft RJ, Louie JL, Herrick JL, Stilwell-Morecraft KS. 2001. Cortical innervation of the facial nucleus in the non-human primate. A new interpretation of the effects of stroke and related subtotal brain trauma on the muscles of facial expression. Brain 124:176–208.

Morecraft RJ, McNeal D, Stilwell-Morecraft KS, Gedney M, Ge J, Schroeder CM, Van Hoesen GW. 2007. Amygdala interconnections with primate cingulate motor cortex: neural correlates for interpreting facial expression and somatomotor signs in temporal lobe epilepsy. J Comp Neurol 500:134–165.

Morecraft RJ, Rockland KS, Van Hoesen GW. 2000. Localization of area prostriata and its projection to the cingulate motor cortex in the rhesus monkey. Cerebral Cortex, 10:192–203.

Morecraft RJ, Van Hoesen GW. 1992. Cingulate input to the primary and supplementary motor cortices in the rhesus monkey: evidence for somatotopy in areas 24c and 23c. J Comp Neurol 322:471–489.

Morecraft RJ, Van Hoesen GW. 1998. Convergence of limbic input to the cingulate motor cortex in the Rhesus monkey. Brain Res Bull 45:209–232.

Morgane PJ, McFarland WL, Jacobs MS. 1982. The limbic lobe of the dolphin brain: a quantitative cytoarchitectonic study. J Hirnforsch. 23:465–552.

Morgane PJ, Galler RJ, Mokler DJ. 2005. A review of systems and networks of the limbic forebrain/limbic midbrain. Prog Neurobiol 75:143–160.

Moruzzi G, Magoun HW. 1949. Brain stem reticular formation and activation of the EEG. Clin Neurophysiol 1:455–473.

Murray RM, Lewis SR. 1987. Is schizophrenia a developmental disorder? Brit Med J 295:681–682.

Naqui NH, Rudrauf D, Damasio H, Bechara A. 2007. Damage to the insula disrupts addiction to cigarette smoking. Science 315:531–534.

Nashold BS, Slaughter DG. 1969. Effects of stimulating or destroying the deep cerebellar regions in man. J Neurosurg 31:172–186.

Nauta WHJ. 1958. Hippocampal projections and realted neural pathways to the midbrain. Brain 81:319–340.

Nauta WHJ, Gygax PA. 1954. Silver impregnation of degenerating axons in the central nervous system: A modified technic. Stain Technol 29:91–93.

Nauta WHJ, Karten HJ. 1970. A general profile of the vertebrate brain, with sidelights on the ancestry of cerebral cortex. In: Schmitt FO, editor. The Neurosciences, second study program. New York: Rockefeller University Press.

Nauta WJ, Domesick VB. 1978. Crossroads of limbic and striatal circuitary: hypothalamo-nigral connections. In: Livingston KE, Hornykiewicz O, editors. Limbic Mechanisms. New Yok: Plenum Press, pp. 75–93.

Nauta WJH. 1986. Ciruitous connections linking cerebral cortex, limbic system and corpus striatum. In: Doane BK, Livingston KE, editors. Limbic System: Functional Organization and Clinical Disorders. New York: Raven Press, pp. 43–54.

Nauta WJH, Feirtag M. 1979. The organization of the brain. Sci Am 241:88–111.

Nauta WJH, Feirtag M. 1986. Fundamental neuroanatomy. New York: W.H. Freeman and Company.

Nauta WJH, Haymaker W. 1969. Hypothalamic nuclei and fiber connections. In: Haymaker W, Anderson E, Nauta WJH (eds.) The Hypothalamus. Springfield, Illinois. Charles C. Thomas, pp. 136–209.

Nauta WJH, Mehler WR. 1966. Projections of the lentiform nucleus in the monkey. Brain Res 1:3–42.

Nauta WJH, Smith GP, Faull RLM, Domesick VB. 1978. Efferent connections and nigral afferents of the nucleus accumbens septi in the rat. Neurosci 3:385–401.

Nestler EJ. 2005. Is there a common molecular pathway for addiction? Nature Neuroscience 8:1445–1449.

Nieuwenhuys R. 1996. The greater limbic system, the emotional motor system and the brain. In: Holstege G, Bandler R, Saper CB, editors. The Emotional Motor System. Prog Brain Res 107:551–580.

Nieuwenhuys R, Meek J. 1990. The telencephalon of actinopterygian fishes. In: Jones EG, Peters A, editors. Cerebral cortex, Vol. 8A. New York: Plenum Press, pp. 31–73.

Nieuwenhuys R, ten Donkelaar HJ, Nicholson C. 1998. The central nervous system of vertebrates. Berlin: Springer.

Nery S, Fishell G, Corbin JG. 2002. The caudal ganglionic eminence is a source of distinct cortical and subcortical cell populations. Nature Neuroscience 5:1279–1287.

Nopoulos PC, Ceilley JW, Gailis EA, Andreasen NC. 1999. An MRI study of cerebellar vermis morphology in patients with schizophrenia; evidence in support of the congnitive dysmetria concept. Biol Psychiatry 46(5):703–711.

Northcutt RG, Kaas JH. 1995. The emergence and evolution of mammalian neocortex. Trend Neurosci 18:373–379.

Oakman SA, Faris PL, Cozzari C, Hartman BK. 1999. Characterization of the extent of pontomesencephalic cholinergic neurons' projections to the thalamus: comparison with projections to midbrain dopaminergic groups. Neurosci 94:529–547.

Oertel WH, Mugnaini E. 1985. Striatal (-like) GABAergic neuronal populations in rat olfactory tubercle, central and medial amygdaloid nuclei. Soc Neurosci Abstr 11:205.

Olszewski. 1957. Reticular formation of the brain. Jasper HH, Proctor LD, Knighton RS, Noshay WC, Costello RT, editors. Little, Brown and Company, Boston, Toronto, p. 56.

Olszewski J, Baxter D. 1954. Cytoarchitecture of the human brain stem. Karger, Basel.

Pandya DN, Kuypers HGJM. 1969. Cortico-cortical connections in the rhesus monkey. Brain Res 13:13–36.

Pandya DN, Yeterian EH. 1985. Architecture and connections of cortical association areas. In: Peters A, Jones EG, editors. Cerebral Cortex, Vol. 4. Plenum, New York.

Panksepp J. 2002. Foreword: The MacLean legacy and some modern trends in emotion research. In: Cory GA, Gardner R, editors. The evolutionary neuroethology of Paul MacLean. Westport: Praeger; pp. ix–xxvii.

Papez JW. 1937. A proposed mechanism of emotion. Arch Neurol Psychiat (Chicago) 42:725–743.

Parkinson JA, Olmstead MC, Burns LH, Robbins TW, Everitt BJ. 1999. Dissociation in effects of lesions of the nucleus accumbens core and shell on appetitive pavlovian approach behavior and the potentiation of conditioned reinforcement and locomotor activity by D-amphetamine. J Neurosci 19:2401–2411.

Parvizi J, Van Hoesen GW, Buckwalter J, Damasio AR. 2006. Neural connections of the posteriomedial cortex in the macaque. Proc Natl Acad Sci USA 103:1563–1568.

Pasquier DA, Anderson C, Forbes WB, Morgane PJ. 1976. Horseradish peroxidase tracing of the lateral habenular-midbrain raphe nuclei connections in the rat. Brain Res Bull 1:443–451.

Pasupathy A, Miller EK. 2005. Different time courses of learning-related activity in the prefrontal cortex and striatum. Nature 433:873–876.

Pearson RCA, Gatter KC, Brodal P, Powell TPS. 1983. The projection of the basal nucleus of Meynert upon the neocortex in the monkey. Brain Research 259:132–136.

Petrides M, Pandya DN. 1984. Projections to the frontal cortex from the posterior parietal region in the rhesus monkey. J Comp Neurol 228:105–116.

Peyron R, Laurent B, Garcia-Larrea L. 2000. Functional imaging of brain responses to pain. Neurophysiology clinics 30:263–288.

Pijnenburg AJJ, Van Rossum JM. 1973. Stimulation of locomotor activity following injection of dopamine into the nucleus accumbens. J Pharm Pharmacol 25:1003–1005.

Ploog DW. 2003. The place of the triune brain in psychiatry. Physiol Behav 79:487–493.

Poloskey SL, Haber S. 2005. Cell proliferation in the striatum during postnatal development. Society for Neuroscience. Ref Type: Abstract.

Price JL. 1995. Thalamus. In: Paxinos G, ed. The rat nervous system, Second Edition, Acdemic Press, San Diego, pp. 629–648.

Price JL. 2004. Olfaction. In: Paxinos G, Mai JK, editors. The human nervous system. Elsevier, Amsterdam, pp. 1197–1211.

Price JL, Carmichael ST, Drevets WC. 1996. Networks related to the orbital and medial prefrontal cortex. A substrate for emotional behavior? In: Holstege G, Bandler R, Saper CB, editors. The emotional motor system program of brain research pp. 461–484.

Price JS. 2002. The triune brain, escalation and de-escalation strategies, and mood disorders. In: Cory GA, Gardner R, editors. The evolutionary neuroethology of Paul MacLean. Westport, Connecticut: Praeger, pp. 107–117.

Puelles L. 2001. Thoughts on the development, structure and evolution of the mammalian and avian telencephalic pallium. Phil Trans R Soc Lond B 356:1–16.

Ramon-Moliner E. 1975. Specialized and generalized dendritic patterns. In: Santini M, editor. Golgi Centennial Symposium Proceedings. New York: Raven Press, pp. 87–100.

Ramon-Moliner E, Nauta WJH. 1966. The isodendritic core of the brainstem. J Comp Neurol 126:311–336.

Rapport LJ, Van Voorhis A, Tzelepis A, Friedman SR. 2001. Executive functioning in adult attention-deficit hyperactivity disorder. Clin Neuropsychol 15:479–491.

Rauch SL, Savage CR, Alpert NM, Fischman AJ, Jenike MA. 1997. The functional neuroanatomy of anxiety: a study of three disorders using PET and symptom provocation. Biol Psychiatry 42:446–452.

Rauch S, Whalen PJ, Shin LM, McInerney SC, Macklin ML, Lasko NB, Orr SP, Pitman RK. 2000. Exaggerated amygdala responses to masked facial stimuli in PTSD, a functional MRI study. Biological Psychiatry 47:769–776.

Rebec GV, Grabner CP, Johnson M, Pierce RC, Bardo MT. 1997a. Transient increases in catecholaminergic activity in medial prefrontal cortex and nucleus accumbens shell during novelty. Neuroscience 76:707–714.

Rebec GV, Christensen JCR, Guerra C, Bardo MT. 1997b. Regional and temporal differences in real-time dopamine efflux in the nucleus accumbens during free-choice novelty. Brain Res 776:61–67.

Reichert KB. 1859. Der Bau des Menschlichen Gehirns durch Abbildung mit erlaut-erndem Texte, 2 Vols. Leipzig: Engelmann.

Reil JC. 1809. Untersuchungen über den Bau des grossen Gehirn im Menschen. Arch Physiol (Halle) 9:136–208.

Reiner A. 1997. An explanation of behavior. Science 250:303–305.

Reinoso BS, Pimenta F, Levitt P. 1996. Expression of the mRNAs encoding the limbic system-associated membrane protein (LAMP): 1. Adult rat brain. J Comp Neurol 375:274–288.

Reynolds SM, Berridge KC. 2001. Fear and feeding in the nucleus accumbens shell: rostrocaudal segregation of GABA-elicited defensive behavior versus eating behavior. J Neurosci 21:3261–3270.

Reynolds SM, Geisler S, Berod A, Zahm DS. 2006. Neurotensin antagonist acutely and robustly attenuates locomotion that accompanies stimulation of a neurotensin-containing pathway from rostrobasal forebrain to the ventral tegmental area. Eur J Neurosci 24:188–196.

Reynolds SM, Zahm DS. 2005. Specificity in the projections of prefrontal and insular cortex to ventral striatopallidum and the extended amygdala. J Neurosci 25:11757–11767.

Ridley RM. 1994. The psychology of perseverative and stereotyped behaviour. Prog Neurobiol 44:221–231.

Risold PY, Swanson LW. 1997a. Chemoarchitecture of the rat lateral septal nucleus. Brain Res Rev 24:91–113.

Robledo P, Robbins TW, Everitt BJ. 1996. Effects of excitotoxic lesions of the central amygdaloid nucleus on the potentiation of reward-related stimuli by intra-accumbens amphetamine. Behav Neurosci 110:981–990.

Rolls ET. 1999. The brain and emotion. Oxford: Oxford University Press.

Rosene DL, Van Hoesen GW. 1987. The hippocampal formation of the primate brain. In: Peters A, Jones EG (Eds) Cerebral Cortex 6, Plenum Press, New York, 345–456.

Rye DB, Saper CB, Lee HJ, Wainer BH. 1987. Pedunculopontine tegmental nucleus of the rat: cytoarchitecture, cytochemistry, and some extrapyramidal connections of the mesopontine tegmentum. J Comp Neurol 259:483–528.

Rye DB, Saper CB, Lee HJ, Wainer BH. 1987. Medullary and spinal efferents of the pedunculopontine tegmental nucleus and adjacent mesopontine tegmentum in the rat. J Comp Neurol 269:315–341.

Rye DB, Wainer BH, Mesualam M-M, Mufson EJ, Saper CB. 1984. Cortical projections arising from the basal forebrain: a study of cholinergic and non-cholinergic components combining retrograde tracing and immunohistochemical localization of choline acetyltransferase. Neuroscience 13:627–643.

Sagan C. 1978. The Dragons of Eden. Random House, New York.

Sakamoto N, Pearson J, Shinoda K, Alheid GF. 1999. In: Bloom FE, Bjorklund A, Hokfelt T (Eds), The Human Basal Forebrain, Part I: An Overview Handbook of Chemical Neuroanatomy, vol. 15. Elsevier, Amsterdam, pp. 1–55.

Salamone JD, Correa M. 2002. Motivational views of reinforcement: implications for understanding the behavioral functions of nucleus accumbens dopamine. Behav Brain Res 137:3–25.

Salamone JD, Cousins MS, Bucher S. 1994. Anhedonia or anergia: effects of haloperidol and nucleus accumbens dopamine depletion on instrumental response selection in a T-maze cost/benefit procedure. Behav Brain Res 65:221–229.

Sanides F. 1957. Die Insulae terminales des Erwachsenen Gehirns des Menschen. J Hirnforsch 3:243–273.

Sanides F. 1958. Vergleichend morphologische Untersuchungen an kleinen Nervenzellen und an Gliazellen. J Hirnforsch 4:113–148.

Sanides F. 1969. Comparative architectonics of the neocortex of mammals and their evolutionary interpretation. Ann NY Acad Sci 167:404–423.

Sano I. 1960. Biochemistry of extrapyramidal motor system. Shinkey Kenkyu no Shinpo (Adv. Neurol. Sci.) 5:42–48. (English translation in: Parkinsonism and Related Disorders 6:3–6, 2000).

Saper CB. 1984. Organization of cerebral cortical afferent systems in the rat. II. Magnocellular basal nucleus. J Comp Neurol 222:313–342.

Sarter M, Bruno JP. 2002. The neglected constituent of the basal forebrain corticopetal projection system: GABAergic projections. Eur J Neurosci 15:1867–1873.

Satoh K, Fibiber HC. 1986. Cholinergic neurons of the laterodorsal tegmental nucleus: efferent and afferent connections. J Comp Neurol 253:277–302.

Scalia F, Halpern M, Knapp H, Riss W. 1968. The efferent connections of the olfactory bulb in the frog: a study of degenerating unmyelinated fibers. J Anat 103:245–262.

Scheibel ME, Scheibel AB. Structural substrates for integrative patterns in the brainstem reticular core. In: Jasper HH, Proctor LD, Knighton RS, Noshay WC, Costello RT, editors. Reticular Formation of the Brain. Boston: Little Brown, pp. 31–55.

Schiller F. 1979. Paul Broca: Founder of French Anthropology, Explorer of the Brain. Berkeley: University of California Press.

Schmahmann JD. 1991. An emerging concept; the cerebellar contribution to higher functions. Arch Neurol 48:1178–1187.

Schmahmann JD. 1997a. Rediscovery of an early concept. In: Schmahmann JD, editor. The cerebellum and cognition. San Diego: Academic Press, pp. 3–27.

Schmahmann JD. 1997b. The cerebellum and cognition. San Diego: Academic Press.

Schneider GE. 1979. Is it really better to have your brain lesion early? A revision of the "Kennard Principle." Neuropsychologia 17:557–583.

Schultz W, Tremblay L, Hollerman JR. 2000. Reward processing in primate orbitofrontal cortex and basal ganglia. Cereb Cortex 10:272–283.

Scoville B, Milner B. 1957. Loss of recent memory after bilateral hippocampal lesions. J Neurol Neurosurg Psychiatry 20:11–21.

Semba K. 1993. Aminergic and cholinergic afferents to REM sleep induction regions of the pontine reticular formation in the rat. J Comp Neurol 330:543–556.

Shammah-Lagnado SJ, Alheid GF, Heimer L. 1999. Afferent Connections of the Interstitial Nucleus of the Posterior Limb of the Anterior Commissure and Adjacent Amygdalostriatal Transition Area in the Rat. Neuroscience In press.

Shammah-Lagnado SJ, Alheid GF, Heimer L. 2001. Striatal and central extended amygdala parts of the interstitial nucleus of the posterior limb of the anterior commissure: evidence from tract-tracing techniques in the rat. J Comp Neurol 439:104–126.

Shammah-Lagnado SJ, Beltramino C, McDonald AJ, Miselis RR, Yang M, Deolmos J, Heimer L, Alheid GF. 2000. The supracapsular bed nucleus of the stria terminlis contains central and medial extended amygdala elements: evidence from anterograde and retrograde tracing experiments in the rat. J Comp Neurol 422:533–555.

Shammah-Lagnado SJ, Santiago AC. 1999. Projections of the amygdalopiriform transition area (APir): A PHA-L study in the rat. Ann NY Acad Sci 655–660.

Sheehan TP, Chambers RA, Russell DS. 2004. Regulation of affect by the lateral seputm: implications for neuropsychiatry. Brain Res Rev 46:71–117.

Sheline YI, Barch DM, Donnelly JM, Ollinger JM, Snyder AZ, Mintun MA. 2001. Increased amygdala response to masked emotional faces in depressed subjects resolves with antidepressant treatment: an fMRI study. Biological Psychiatry 50: 651–658.

Shelley BP, Trimble MR. 2004. The insular Lobe of Reil—its anatomico-functional, behavioural and neuropsychiatric attributes in humans—a review. World Journal of Biological Psychiatry 5:176–200.

Shi CJ, Cassell MD. 1998. Cortical, thalamic, and amygdaloid connections of the anterior and posterior insular cortices. J Comp Neurol 399:440–468.

Shinnick-Gallagher P, Pitkänen A, Shekhar A, Cahill L. 2003. The amygdala in brain function. Basic and clinical approaches. New York: New York Academy of Sciences.

Shreve PE, Uretsky NJ, 1988. Effect of GABAergic transmission in the subpallidal region on the hypermotility response to the administration of excitatory amino acids and picrotoxin into the nucleus accumbens. Neuropharmacol 27:1271–1275.

Shute CCD, Lewis PR. 1967. The ascending cholinergic reticular system: neocortical olfactory and subcortical projections. Brain 90:497–520.

Silfenius H, Gloor P, Rasmussen T. 1964. Evaluation of insular ablation in surgical treatment of temporal lobe epilepsy. Epilepsia 5:307–320.

Smeets WJAJ. 1990. The telencephalon of cartilaginous fishes. In: Jones EG, Peters A, editors. Cerebral cortex, Vol. 8A. New York: Plenum Press, pp. 3–30.

Smiley JF, Mesulam M-M. 1999. Cholinergic neurons of the nucleus basalis of Meynert receive cholinergic, catecholaminergic and GABAergic synapses: and electron microscopic investigation in the monkey. Neurosci 88:241–255.

Sobel N, Prabhakaran V, Desmond JE, et al. 1998. Sniffing and smelling, separate subsystems in the human olfactory cortex. Nature, 392:282–286.

Solodkin A, Van Hoesen GW. 1996. Entorhinal cortex modules of the human brain. J Comp Neurol 365:610–627.

Somogyi P, Bolam JP, Totterdell S, Smith AD. 1981. Monosynaptic input from the nucleus accumbens-ventral striatal region to retrogradely labelled nigrostriatal neurones. Brain Res 217:245–263.

Song DD, Harlan RE. 1993. Ontogeny of the proenkephalin system in the rat corpus striatum: its relationship to dopaminergic innervation and transient compartmental expression. Neuroscience 52:883–909.

Song DD, Harlan RE. 1994. The development of enkephalin and substance P neurons in the basal ganglia: insights into neostriatal compartments and the extended amygdala. Brain Res Developmental Brain, 247–261.

Sorensen KE, Witter MP. 1983. Entorhinal efferents reach the caudate-putamen. Neurosci Letters, 35:259–264.

Sternberger LA, Cuculis JJ. 1969. Method for enzymatic intensification of the immunocytochemical reaction without use of labeled antibodies. J Histochem Cytochem 17:190.

Stevens JR. 1973. An anatomy of schizophrenia? Arch Gen Psychiatry 29:177–189.

Stevens JR. 1992. Abnormal reinnervation as a basis for schizophrenia: a hypothesis. Arch Gen Psychiatry 49:238–243.

Stevens JR. 2002. Schizophrenia: reproductive hormones and the brain. Am J Psychiatry 159:713–719.

Strakowski SM, DelBello MP, Sax KW, Zimmerman ME, Shear PK, Hawkins JM, Larson ER. 1999. Brain magnetic resonance imaging of structural abnormalities in bipolar disorder. Arch Gen Psychiat, 56:254–260.

Strenge H, Braak E, Braak H. 1977. Über den Nucleus striae terminalis im Gehirn des erwachsenen Menschen. Z Mikrosk Anat Forsch 91:1.S.:105–118.

Suzuki WA, Amaral DG. 1994. Topographic organization of the reciprocal connections between the monkey entorhinal cortex and the perirhinal and parahippocampal cortices. J Neurosci 14:1856–1877.

Swanson LW. 1976. An autoradiographic study of the efferent connections of the preoptic region in the rat. J Comp Neurol 167:227–256.

Swanson LW. 1987. Limbic system. In: Adelman G, Smith BH, editors. Encyclopedia of Neuroscience. 2nd ed. Vol. 2, Amsterdam: Elsevier; pp. 1053–1055.

Swanson LW. 1987. The Hypothalamus. In: Björkland A, Hökfelt T, Swanson LW, editors. Handbook of Chemical Neuroanatomy. Volume 5. Integrated Systems of the CNS, Part I. Hypothalamus, Hippocampus, Amygdala, Retina. Amsterdam: Elsevier, pp. 125–277.

Swanson LW. 2000. Cerebral hemisphere regulation of motivated behavior. Brain Res 886:113–164.

Swanson LW. 2003. Brain architecture. Oxford: Oxford University Press.

Swanson LW. 2003. The amgydala and its place in the cerebral hemisphere. Ann NY Acad Sci 985:174–185.

Swanson LW. 2003. The amygdala and its place in the cerebral hemisphere. In: Shinnick-Gallagher P, Pitkänen A, Shekhar A, and Cahill L, editors. The amygdala in brain function. New York: New York Academy of Sciences, pp. 174–184.

Swanson LW, Cowan WM. 1975. A note on the connections and development of the nucleus accumbens. Brain Res 92:324–330.

Swanson LW, Köhler C, Björkland A. The limbic region. I. The septohippocampal system. In: Björkland A, Hökfelt T, Swanson LW, editors. Handbook of Chemical Neuroanatomy. Volume 5. Integrated Systems of the CNS, Part I. Hypothalamus, Hippocampus, Amygdala, Retina. Amsterdam: Elsevier; 1987, pp. 125–277.

Swanson LW, Mogenson GJ, Gerfen CR, Robinson P. 1984. Evidence for a projection from the lateral preoptic area and substantia innominata to the "mesencephalic locomotor region" in the rat. Brain Research 295:161–178.

Swanson LW, Mogenson GJ, Simerly RB, Wu M. 1987. Anatomical and electrophysiological evidence for a projection from the medial preoptic area to the "mesencephalic and subthalamic locomotor regions" in the rat. Brain Res 405:108–122.

Swanson LW, Petrovich GD. 1998. What is the amygdala? Trends Neurosci 21: 323–331.

Switzer RC, Hill J, Heimer L. 1982. The globus pallidus and its rostroventral extension into the olfactory tubercle of the rat: a cyto- and chemoarchitectural study. Neuroscience 7:1891–1904.

Talbot K, Woolf NJ, Butcher LL. 1988a. Feline islands of Calleja complex: 11. Cholinergic and cholinesterasic features. J Comp Neurol 275:580–603.

Talbot K, Woolf NJ, Butcher LL. 1988b. Feline Islands of Calleja complex: I. Cytoarchitectural organization and comparative anatomy. J Comp Neurol 275:553–579.

Tamminga CA, Thaker GK, Buchanan R, Kirkpatrick B, et al. 1992. Limbic system abnormalities identified in schizophrenia using PET with fluorodeoxyglucose. Archives of General Psychiatry 49:522–530.

Tillfors M, Furmark T, Marteinsdottir I, Fischer H, Pissiota A, Langstrom B, Fredrikson M, et al. 2001. Cerebral blood flow in subjects with social phobia during stressful speaking tasks: a PET study. Am J Psychiat, 158:1220–1226.

Todtenkopf MS, Stellar JR, Williams EA, Zahm DS. 2004. Differential distribution of parvalbumin immunoreactive neurons in the striatum of cocaine sensitized rats. Neurosci 127:35–42.

Trimble MR. 1991. The Psychoses of Epilepsy. Raven Press, New York.

Trimble MR. 1996. Biological Psychiatry, 2nd edition, Chichester, J Wiley and sons.

Trimble MR. 2007. The soul in the brain. Johns Hopkins Press, Baltimore.

Trimble MR, Tebartz van Elst L. 1999. On Some Clinical Implications of the Ventral Striatum and the Extended Amygdala. In: Advancing from the Ventral Striatum to the Extended Amygdala. Ann New York Acad Sci 877:638–644.

Trimble MR, Zarifian E. 1984. Psychopharmacology of the limbic system. Oxford: Oxford University Press.

Turner BH, Gupta KC, Mishkin M. 1978. The locus and cytoarchitecture of the projection areas of the olfactory bulk in Macaca mulatta. J Comp Neurol 177:381–396.

Turner BJ, Zimmer J. 1984. The architecture and some of the interconnections of the rat's amygdala and lateral periallocortex. J Comp Neurol 227:540–557.

Turetsky BI, Moberg PJ, Roalf DR, Arnold SE, Gur R. 2003. Decrements in volume of the Anterior Ventromedial Temporal Lobe and Olfactory Dysfunction in Schizophrenia. Archives of General Psychiatry 60:1193–1200.

Usuda I, Tanaka K, Chiba T. 1998. Efferent projections of the nucleus accumbens in the rat with special reference to subdivision of the nucleus: biotinylated dextran amine study. Brain Res 797:73–93.

Vanderwolf CH, Kolb B, Cooley RK. 1978. Behavior of the rat after removal of the neocortex and hippocampal formation. J Comp Physiol Psych 92:156–175.

Van Hoesen GW. 1981. The differential distribution diversity and sprouting of cortical projections to the amygdala in the rhesus monkey. In: Ben Ari Y, editor. The Amygdaloid Complex, Elsevier, Amsterdam, 77–90.

Van Hoesen GW. 1982. The parahippocampal gyrus: New observations regarding its cortical connections in the monkey. Trend Neurosci 5:345–350.

Van Hoesen GW. 1997. Ventromedial temporal lobe anatomy, with comments on Alzheimer's disease and temporal injury. In: Salloway S, Malloy P, Cummings JF

(Eds.) The Neuropsychiatry of Limbic and Subcortical Disorders, American Psychiatric Press, Inc. Washington, D.C. pp. 19–29.

Van Hoesen GW. 2002. The human parahippocampal region in Alzheimer's disease, dementia, and aging. In: Witter M, Wouterlood F (Eds.) The Parahippocampal Region, Oxford, New York.

Van Hoesen GW, Hyman BT, Damasio AR. 1991. Entorhinal cortex pathology in Alzheimer's disease. Hippocampus 1:1–18.

Van Hoesen GW, Mesulam MM, Haaxma R. 1976. Temporal cortical projections to the olfactory tubercle in the rhesus monkey. Brain Res 109:375–381.

Van Hoesen GW, Morecraft RJ, Vogt BA. 1993. Connections of the monkey cingulate cortex. In: Vogt BA, Gabriel M (Eds.) The Neurobiology of Cingulate Cortex, Limbic Thalamus, Birkhauser, Boston, 249–284.

Van Hoesen GW, Pandya DN, Butters N. 1972. Cortical afferents to the entorhinal cortex of the rhesus monkey. Science 175:1471–1473.

Van Hoesen GW, Solodkin A. 1993. Some modular features of temporal cortex in humans as revealed by pathological changes in Alzheimer's disease. Cereb Cortex 3:465–475.

Varma A, Trimble MR. 1997. Subcortical neurological syndromes. In: Trimble MR, Cummings J, eds. Contemporary Behavioural Neurology. Newton, Butterworth Heinemann, pp. 239–254.

Verh K. 2003. Deep brain stimulation in treatment refractory obsessive compulsive disorder. Acad Geneeskd Belg 65:385–399.

Vogt BA, Laureys S. 2005. Posterior cingulate, precuneal and retrosplenial cortices: cytology and components of the neural network correlates of consciousness. Progress in Brain Research, 150:205–217.

Vogt BA, Pandya DN. 1987. Cingulate cortex of the rhesus monkey: II. Cortical afferents. J Comp Neurol 262:271–289.

Vogt BA, Vogt L, Faber NB, Bush G. 2005. Architecture and neurocytology of monkey cingulate gyrus. J Comp Neurol 485:218–239.

Vogt BA, Vogt L, Laureys S. 2006. Cytology and functionally correlated circuits of the human posterior cingulate areas. Neuroimage 29:452–466.

Vogt C, Vogt O. 1922. Erkrankungen der Grosshirnrinde im Lichte der Topistik, Pathoklise und Pathoarchitektonik, J Psychol Neurol (Lpz) 28:9–171.

Vogt O. 1925. Der Begriff der Pathoklise. J Psychol Neurol (Lpz) 31:245–255.

Volkow ND, Li. 2004. Drug addiction: the neurobiology of behavior gone awry. Nat Rev Neurosci 5:963–970.

Voorn P, Brady LS, Berendse HW, Richfield EK. 1996. Densitometrical analysis of opioid receptor ligand binding in the human striatum I: Distribution of μ opioid receptor defines shell and core of the ventral striatum. Neuroscience 75:777–792.

Walker DL, Toufexis DJ, Davis M. 2003. Role of the bed nucleus of the stria terminalis versus the amygdala in fear, stress, and anxiety. Eur J Pharmacol 463:199–216.

Weinberger DR. 1987. Implications of normal brain development for the pathogenesis of schizophrenia. Arch Gen Psychiatr 44:660–669.

Weiskrantz L. 1956. Behavioral changes associated with ablation of the amygdaloid complex in monkeys. Comp Physiol Psychol 49:381–391.

West AR, Grace AA. 2001. The role of frontal-subcortical circuits in the pathophysiology of schizophrenia. In: Lichter DG and Cummings J, editors. Frontal-subcortical circuits in psychiatric and neurological disorders. New York: The Guilford Press. pp. 372–400.

Whishaw I. 1990. The decorticate rat. In: Kolb B, Tees RC, editors. The cerebral cortex of the rat. MIT Press, Cambridge MA, pp. 239–267.

Willis T. 1664. Cerebri Anatomie. London: Martzer and Alleftry.

Winn P. 1998. Frontal syndrome as a consequence of lesions in the pedunculopontine tegmental nucleus: a short theoretical review. Brain Res Bull 47:551–563.

Winn P. 2006. How best to consider the structure and function of the pedunculopontine tegmental nucleus: Evidence from animal studies. Journal of the Neurological Sciences 248:234–250.

Wise RA. 1978. Catecholamine theories of reward: a critical review. Brain Res 52:215–247.

Wise RA. 1980. Action of drugs of abuse on brain reward systems. Pharmacol Biochem Behav 13 Suppl 1:213–223.

Wise RA. 2004. Dopamine, learning and motivation. Nature Reviews Neuroscience 5:1–12.

Wise RA, Rompré P-P. 1989. Brain dopamine and reward. Ann Rev Psychol 40:191–225.

Witter MF. 1993. Organization of the entorhinal-hippocampal system: a review of current anatomical data. Hippocampus 3:33–44.

Witter M, Groenewegen HJ, Lopes Da Silva FH, Lohman AHM. 1989. Functional organization of the extrinsic and intrinsic circuitry of the parahippocampal region. Prog Neurobiol 33:161–254.

Woodruff GN, Kelly PH, Elkhawad AO. 1976. Effects of dopamine receptor stimulants on locomotor activity of rats with electrolytic or 6-hydroxydopamine-induced lesions of the nucleus accumbens. Psychopharmacol 46:195–198.

Woods JW. 1964. Behavior of chronic decerebrate rats. J Neurophysiol 27:635–644.

Woolf NJ, Butcher LL. 1986. Cholinergic systems in the rat brain: III. Projections from the pontomesencephalic tegmentum to the thalamus, tectum, basal ganglia and basal forebrain. Brain Res Bull 16:603–637.

Woolf NJ, Eckenstein F, Butcher LL. 1983. Cholinergic projections from the basal forebrain to the frontal cortex: a combined fluorescent tracer and immunohistochemical analysis in the rat. Neurosci Lett 40:93–98.

Woolf NJ, Eckenstein F, Butcher LL. 1984. Cholinergic systems in the rat brain. I. Projections to the limbic telencephalon. Brain Res Bull 13:751–784.

Woolf NJ, Hernit MC, Butcher LL. 1986. Cholinergic and non-cholinergic projections from the rat basal forebrain revealed by combined choline acetyltransferase and Phaseolus vulgaris-leucoagglutinin immunohistochemistry. Neurosci Letts 66:281–286.

Yakovlev PI. 1948. Motility, behavior, and the brain. J Nerv Ment Dis 107:313–335.

Yakovlev PI. 1959. Pathoarchitectonic studies of cerebral malformations. J Neuropath Exper Neurol 18:22–55.

Yakovlev PI. 1972. A proposed definition of the limbic system. In: Hockman CH, editor. Limbic System Mechanisms and Automatic Function, Charles C Thomas, Springfield, IL, pp. 241–283.

Yasargil MG. 1994. Microneurosurgery, vol. 4, Thieme Medical Publishers, Inc., Stuttgart.

Yeterian EH, Van Hoesen GW. 2004. Cortico-striate projections in the rhesus monkey: the organization of certain cortico-caudate connections. Brain Res 139:43–63.

Young WS III, Alheid GF, Heimer L. 1984. The ventral pallidal projection to the mediodorsal thalamus: a study with fluorescent retrograde tracers and immunohistofluorescence. J Neurosci 4:1626–1638.

Záborszky L. 2002. The modular organization of brain systems. Basal forebrain: the last frontier. In: Azmitia EC, DeFelipe J, Jones EG, Rakic P, Ribak CE, editors. changing views of Cajal's neuron. Amsterdam: Elsevier; Prog Brain Res 136: 359–372.

Záborszky L, Alheid GF, Beinfeld MC, Eiden LE, Heimer L, Palkovits M. 1985. Cholecystokinin innervation of the ventral striatum: a morphological and radioimmunological study. Neuroscience 14:427–453.

Záborszky L, Carlsen J, Brashear HR, Heimer L. 1986. Cholinergic and GABAergic afferents to the olfactory bulb in the rat with special emphasis on the projection neurons in the nuclous of the horizontal limb of the diagonal band. J Comp Neurol 243:488–509.

Záborszky L, Cullinan WE. 1992. Projections from the nucleus accumbens to cholinergic neurons of the ventral pallidum: a correlated light and electron microscopic double-immunolabeling study. Brain Res 570:92–110.

Záborszky L, Cullinan WE, Braun A. 1991. Afferents to basal forebrain cholinergic neurons: an update, In: Napier TC, Kalivas PW, Hanin I, editors. The basal forebrain. New York: Plenum Press, pp. 43–100.

Záborszky L, Pang K, Somogyi J, Nadasdy Z, Kallo I. 1999. The basal forebrain corticopetal system revisited. Ann NY Acad Sci 877:339–367.

Záborszky L, Wouterlood FG, Lanciego JL. 2006. Neuroanatomical tract-tracing 3; molecules, neurons and systems. New York: Springer.

Zahm DS. 1998. Is the caudomedial shell of the nucleus accumbens part of the extended amygdala? A consideration of connections. Crit Rev Neurobiol 12:245–265.

Zahm DS. 1999. Functional-anatomical implications of nucleus accumbens core and shell subterritories. Ann NY Acad Sci 877:113–129.

Zahm DS. 2000. An integrative neuroanatomical perspective on some subcortical substrates of adaptive responding with emphasis on the nucleus accumbens. Neurosci Biobehav Rev 24:85–105.

Zahm DS. 2006. The evolving theory of basal forebrain functional-anatomical "macrosystems." Neurosci Biobehav Rev 30:148–172.

Zahm DS, Brog JS. 1992. On the significance of subterritories in the "accumbens" part of the rat ventral striatum. Neuroscience 50:751–767.

Zahm DS, Grosu S, Irving JC, Williams EA. 2003. Discrimination of striatopallidum and extended amygdala in the rat: a role for parvalbumin immunoreactive neurons. Brain Res 978:141–154.

Zahm DS, Heimer L. 1987. The ventral striatopallidothalamic projection. III. Striatal cells of the olfactory tubercle establish direct synaptic contact with ventral pallidal cells projecting to mediodorsal thalamus. Brain Res 404:327–331.

Zahm DS, Heimer L. 1993. The efferent projections of the rostral pole of the nucleus accumbens in the rat: Comparison with the core and shell projection patterns. J Comp Neurol 327:220–232.

Zahm DS, Jensen S, Williams EA, Martin JR III. 1999. Direct comparison of projections from the central nucleus of the amygdala and nucleus accumbens shell. Eur J Neurosci 11:1119–1126.

Zahm DS, Vogel AC, Porter A, Williams K, Loopuijt LD. Some properties and relationships of a caudal subpopulation of basal forebrain cholinergic neurons in the rat. Soc Neurosci Abstr 2004; Prog # 754.18.

Zahm DS, Williams EA, Krause JE, Welch MA, Grosu DS. 1998. Distinct and interactive effects of d-amphetamine and haloperidol on levels of neurotensin and its mRNA in subterritories in the dorsal and ventral striatum of the rat. J Comp Neurol 400:487–503.

Zahm DS, Williams EA, Latimer MP, Winn P. 2001. Ventral mesopontine projections of the caudomedial shell of the nucleus accumbens and extended amygdala in the rat: Double dissociation by organization and development. J Comp Neurol 436:111–125.

Zahm DS, Williams EA, Wohltmann C. 1996. The ventral striatopallidothalamic projection: IV. Relative contributions from neurochemically distinct pallidal subterritories in the subcommissural region and olfactory tubercle and from adjacent extrapallidal parts of the rostrobasal forebrain. J Comp Neurol 364:340–362.

Zahm DS, Záborszky L, Alheid GF, Heimer L. 1987. The ventral striatopallidothalamic projection: II. The ventral pallidothalamic link. J Comp Neurol 255:592–605.

Zald DH. 2003. The human amygdala and the emotional evaluation of sensory stimuli. Brain Res Brain Res Rev 41:88–123.

Ziehen T. 1897. Das Zentralnervensystem der Monotremen und Marsupialier, I. II. Denkschriften der Medizinischnaturwissenschaftlichen Gesellschaft zu Jena Bd. 6.

Zigmond MJ, Bloom FE, Lanis SC, Roberts JL, Squire LR. 1999. Fundamental neuroscience. San Diego: Academic Press.

Zilles K. 1990. Cortex. In Paxinos G (Ed.). The Human Nervous System. New York. Academic Press. pp. 757–802.

INDEX

Page numbers followed by "f" denote figures

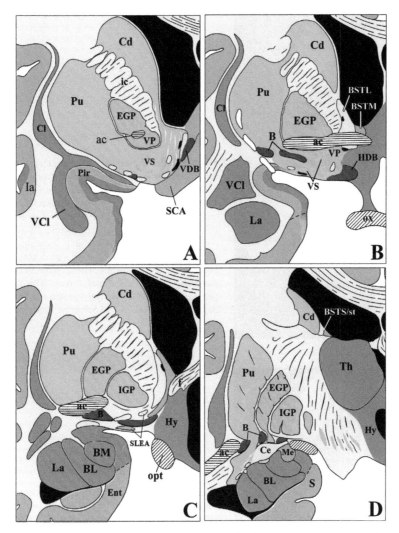

FIGURE 3.1 Schematic drawings showing the human basal forebrain in a series of coronal sections starting rostrally at the level of the accumbens in A and ending at the level of the caudal amygdala in D. Color coding: striatum—light blue; pallidum—salmon; extended amygdala central division—yellow; extended amygdala medial divsion—green; basal forebrain magnocellular complex—brown; primary olfactory areas—pink; Abbreviations: ac—anterior commissure; B—basal nucleus of Meynert; BL—basolateral amygdala; BM—basomedial amygdala; BSTL and BSTM—lateral and medial divisions of the bed nucleus of the stria terminalis; BSTS/st—supracapsular/stria terminalis division of the extended amygdala (shown in yellow); Cd—caudate nucleus; Ce—central nucleus of the amygdala; Cl—claustrum; EGP—external segment of the globus pallidus; Ent—entorhinal cortex; f—fornix; Hy—hypothalamus; IGP—internal segment of the globus pallidus; La—lateral amygdala; Me—medial nucleus of the amygdala; opt—optic tract; ox—optic chiasm; Pir—piriform (primary olfactory) cortex; Pu—putamen; S—subiculum; SCA—subcallosal area; SLEA—sublenticular extended amygdala (shown in yellow [central division] and green [medial division]); Th—thalamus; VCl—ventral claustrum; VDB—vertical limb of the diagonal band of Broca; VP—ventral pallidum; VS—ventral striatum. (Art by Medical and Scientific Illustration, Crozet, Virginia; reprinted from Heimer et al., 1999 with permission.)

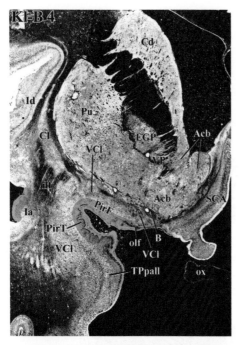

FIGURE 3.2 Klüver-Barrera stained frontal section through the human brain at the level of the optic chiasm. Primary olfactory cortex is highlighted in pink. Abbreviations: Acb—accumbens; B—basal nucleus of Meynert; Cd—caudate nucleus; Cl—claustrum; EGP—external segment of the globus pallidus; Id—dorsal insular cortex; Ia—anterior insular cortex; ilf—inferior longitudinal fasciculus; olf—olfactory tract; ox—optic chiasm; PirF—piriform cortex, frontal portion; PirT—piriform cortex, temporal portion; Pu—putamen; SCA—subcallosal area; TPpall—temporopolar periallocortex; VCl—ventral claustrum; VP—ventral pallidum. (Art by Medical and Scientific Illustration, Crozet, Virginia; reprinted from Sakamoto et al., 1999 with permission.)

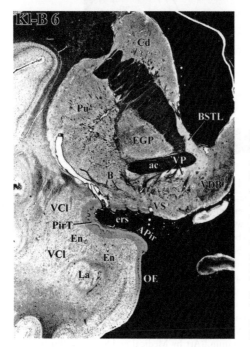

FIGURE 3.3 Klüver-Barrera stained frontal section through the human brain at a level just rostral to the crossing of the anterior commissure. Primary olfactory cortex is highlighted in pink, central division of the extended amygdala in yellow. Abbreviations: ac—anterior commissure; Acb—accumbens; APir—amygdalopiriform transition cortex; B—basal nucleus of Meynert; BSTL—lateral division of the bed nucleus of the stria terminalis; Cd—caudate nucleus; Cl—claustrum; EGP—external segment of the globus pallidus; En—endopiriform nucleus; ers—endorhinal sulcus; LA—lateral amygdala; PirT—piriform cortex, temoral portion; Pu—putamen; TPpall—temporopolar periallocortex; VCl—ventral claustrum; VDB—vertical limb of the diagonal band of Broca; VP—ventral pallidum; VS—ventral striatum. (Art by Medical and Scientific Illustration, Crozet, Virginia; reprinted from Sakamoto et al., 1999 with permission.)

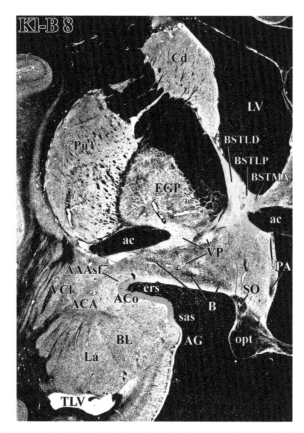

FIGURE 3.4 Klüver-Barrera stained frontal section through the human brain at the level of the crossing of the anterior commissure. Olfactory bulb projections to the cortical amygdaloid nucleus are highlighted in pink, central and medial divisions of the extended amygdala in yellow and green, respectively. The large hyperchromatic cells of the basal nucleus of Meynert (B) form particularly large conglomerates at this level. Abbreviations: AAAsf—anterior amygdaloid area, superficial division; ac—anterior commissure; ACA—amygdaloclaustral area; Acb—accumbens; Aco—cortical nucleus of the amygdala; AG—ambiens gyrus; APir—amygdalopiriform transition cortex; B—basal nucleus of Meynert; BL—basolateral thalamic nucleus; BSTMA, BSTLD, and BSTLP—medial division, anterior part and lateral division, dorsal and posterior parts of the bed nucleus of the stria terminalis; Cd—caudate nucleus; ers—endorhinal sulcus; LA—lateral amygdala; LV—lateral ventricle; opt—optic tract; Pu—putamen; SO—supraoptic nucleus; TLV—temporal horn of the lateral ventricle; Pa—hypothalamic paraventricular nucleus; sas—semiannular sulcus; TPpall—temporopolar periallocortex; VCl—ventral claustrum; VDB—vertical limb of the diagonal band of Broca; VP—ventral pallidum; VS—ventral striatum. (Art by Medical and Scientific Illustration, Crozet, Virginia; reprinted from Sakamoto et al., 1999 with permission.)

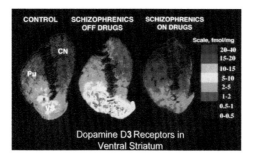

CBOX 4 FIGURE A PET imaging demonstrating elevated density of dopamine D3 receptors in the ventral striatum of schizophrenics when off antipsychotic drugs as compared to controls is reduced by antipsychotic drug treatment. (Reprinted from Gurevich et al., 1997, with permission.)

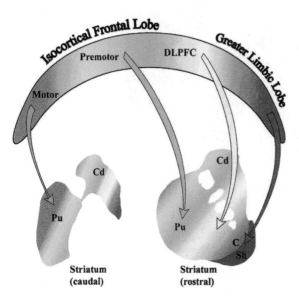

FIGURE 3.8 Diagram of corticofugal projections from the isocortical frontal lobe and the greater limbic lobe of the human brain. The gradual changes in color indicate overlapping projections. Although the crescent-shaped figure represents the isocortical frontal lobe and the nonisocortical greater limbic lobe only, it should be emphasized that cortical-striatal projections do come from the entire cerebral cortex. The greater limbic lobe regions include all nonisocortical parts of the cerebral cortex, including the cortical-like laterobasal-cortical amygdala (see Figs. 4.1 and 4.4). (Original graphic artwork by Suzanne Haber, modified from Heimer, 2003, with permission.)

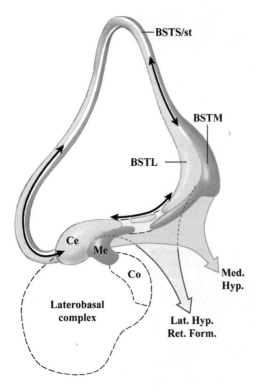

FIGURE 3.10 The extended amygdala (yellow and green) shown in isolation from the rest of the brain, with the extensions of the central (Ce) and medial (Me) amygdaloid nuclei within the stria terminalis (st) and through the sublenticular region to the bed nucleus of stria terminalis (BST). The central division of the extended amygdala is coded in yellow and the medial division in green and the major outputs from each are indicated by the respective yellow and green arrows. The black arrows represent the abundant long and short intrinsic associational axonal connections that characterize the macrosystems. Note that the laterobasal-cortical amygdaloid complex, which is included in the concept of the limbic lobe, is not part of the extended amygdala. Further abbreviations: BSTL—lateral bed nucleus of the stria terminalis; BSTM—medial bed nucleus of the stria terminalis; BSTS/st—supracapsular part or the bed nucleus of stria terminalis; Co—cortical amygdaloid nucleus. (Modified from Heimer et al., 1999, with permission; art by Medical and Scientific Illustration, Crozet, VA.)

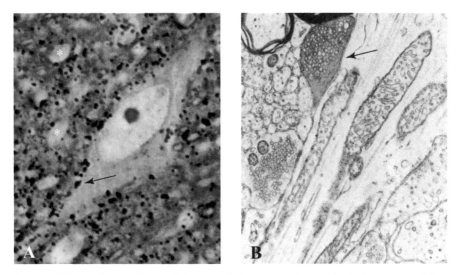

BSBOX 2 FIGURE D High-magnification light (A) and electron (B) micrographs depicting terminal degeneration in plastic-embedded sections of the rat brain. The section in A was stained with the Fink-Heimer technique two days following the creation of a lesion similar to the one shown in BSB Fig C. It shows terminal degeneration at the site marked with an arrow in both panels in BSB 2 Fig C. The section in B illustrates the same site, also from a similarly lesioned brain, processed for electron microscopy. The structure indicated by the arrow is a striatopallidal bouton showing increased electron density and swollen vesicles indicative of terminal degeneration. (Reprinted from Heimer, 2003, with permission.)

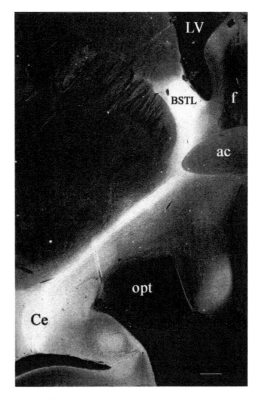

BSBOX 3 FIGURE B Autoradiography of amino acid transport after an injection in the central nucleus (Ce) of the macaque amygdala (dark-field illumination). Note the dense labeling of axons and terminals in the sublenticular extended amygdala. This high degree of associative connections distinguishes the extended amygdala from the overlying striatopallidal complex. Abbrevations: ac—anterior commissure; BST—bed nucleus of the stria terminalis; GP—globus pallidus; opt—optic tract. (Reprinted from Alheid et al., 1990, with permission.)

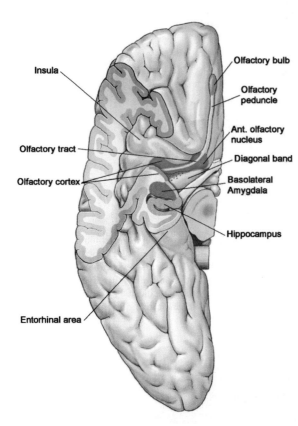

Labels on figure:

Insula

Olfactory bulb

Olfactory peduncle

Ant. olfactory nucleus

Olfactory tract

Diagonal band

Olfactory cortex

Basolateral Amygdala

Hippocampus

Entorhinal area

FIGURE 4.3 Ventral view of the human brain with the limbic lobe depicted in four colors. Darker green depicts the olfactory bulb and the location of the frontal and temporal primary olfactory allocortex. Lighter green depicts orbitofrontal, insular, and parahippocampal agranular, and dysgranular and periallocortical regions also belonging to the limbic lobe. Frontal opercular and temporal polar areas are dissected away to appreciate these and the location of the amygdala (orange) and hippocampus (violet). Reprinted from Heimer and Van Hoesen (2006) with permission. (Artwork provided by Medical and Scientific Illustration, Crozet, VA, USA.)

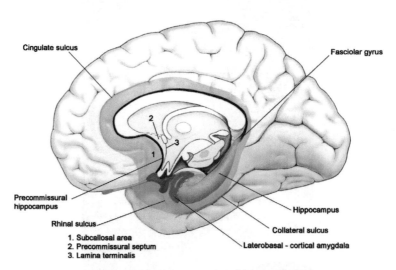

Labels on figure:

Cingulate sulcus

Fasciolar gyrus

Precommissural hippocampus

Rhinal sulcus

1. Subcallosal area
2. Precommissural septum
3. Lamina terminalis

Hippocampus

Collateral sulcus

Laterobasal - cortical amygdala

FIGURE 4.4 Medial views of the human brain depicting the components of the limbic lobe in colors. Note the inner ring of allocortices in dark green, light purple, and dark purple, and the outer ring in light green, denoting the agranular, dysgranular, and dyslaminate cortices. Reprinted from Heimer and Van Hoesen (2006) with permission. (Artwork provided by Medical and Scientific Illustration, Crozet, VA, USA.)

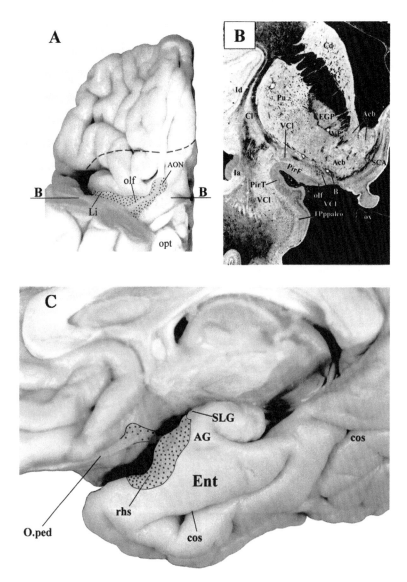

FIGURE 4.6 A–C: A is a photograph of the human orbital cortex following removal of the temporal pole. The interrupted line denotes the approximate anterior border of the limbic lobe, and the magenta-colored area denotes the projection zone of the olfactory bulb. The line labeled B-B denotes the line of cut for the coronal section shown in photograph B. As in A, the magenta color denotes the terminal zone for olfactory bulb projections. Photograph C shows the same, but it is mapped onto a ventromedial view of the human brain. Direct olfactory bulb projections extend beyond the temporal and frontal olfactory piriform allocortex and end in peri-olfactory agranular cortices in the insula, orbitofrontal, temporal polar, and parahippocampal regions. The latter includes area EO, the most anterior subdivision of the entorhinal periallocortex. Abbreviations: Acb—accumbens; AG—gyrus ambiens; AON—anterior olfactory nucleus; B—basal nucleus of Meynert; Cd—caudate nucleus; CL—claustrum; cos—collateral sulcus; EGP—globus pallidus; Ent—entorhinal cortex; Ia—agranular insula; Id—dysgranular insula; Li—limen insulae, olf—olfactory tract; O.ped—olfactory peduncle; opt—optic tract; ox—optic chiasm; PirE—frontal piriform cortex; Pirt—temporal piriform cortex; Pu—putamen; SCA—subcallosal gyrus; rhs—rhinal sulcus; SLG—semilunar gyrus; VCL—ventral claustrum. Note the abbreviation TPppaleo identifies Brodmann's area 35, the perirhinal cortex, a type of peri-olfactory cortex described in the text. Reprinted from Heimer and Van Hoesen (2006) with permission.

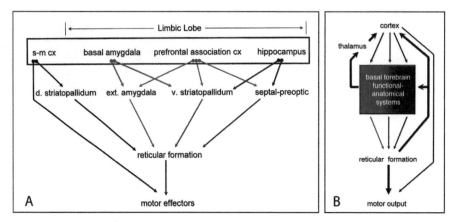

FIGURE 5.2 Connectional relationships of basal forebrain macrosystems. A. Outputs from greater limbic lobe structures, including the basal amygdala, prefrontal cortex (cx), and hippocampus diverge to innervate multiple macrosystems, including extended (ext.) amygdala, ventral (v.) striatopallidum, and septal-preoptic system, which, together with the dorsal striatopallidum, give rise to outputs that converge in the reticular formation. Outputs from the reticular formation and cortex, in turn, converge upon motor effectors. B. Replicates features illustrated in A, but, in addition, emphasizes reentrant pathways returning (1) to the cortex via the thalamus and (2) to the cortex and/or deep telencephalic nuclei (the macrosystems themselves) via ascending modulatory projection systems. Additional abbreviation: s-m—sensorimotor.

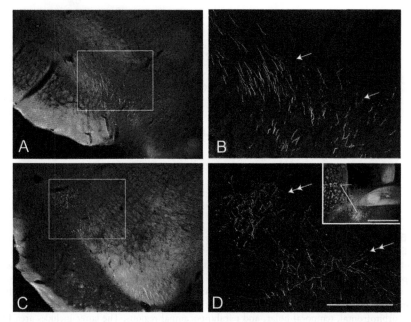

FIGURE 5.6 Sections through the ventral tegmental area (A, B) and rostral mespontine tegmentum (C and D), showing that labeled axonal projections originating in the central division of the extended amygdala pass throught the ventral tegmental area and medial substantia nigra lacking varicosities and thus are likely to be fibers of passage (single arrows). In contrast, these projections exhibit a robust burst of axonal varicosities indicative of functional synaptes in the retrorubral field, particularly laterally (double arrows in D). Boxes in A and C are enlarged as B and D, respectively. The inset in D shows the PHA-L injection site in the bed nucleus of stria terminalis (BST) and also shows the robust associational connections within the extended amygdala (indicated by lines). Scale bar in D is 1.0 and 0.4mm for A, C, and B, D, respectively. Bar in inset is 1.0mm. Modified from Zahm, 2006, with permission.

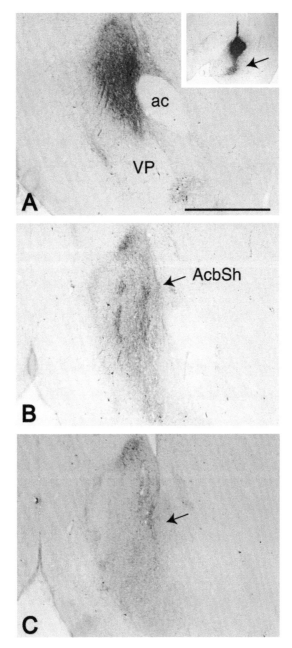

BSBOX 8 FIGURE A Micrographs illustrating axonal transport following an injection of bidirectionally transported cholera toxin β subunit (CTβ) into the central nucleus of the amygdala (CeA, the injection site is shown in the inset in A). In contrast to PHA-L injections into the CeA, which produce negligible anterograde labeling in the nucleus accumbens shell (AcbSh, see BSB 8 Fig. B), this injection produced robust labeling there (arrows in panels B and C; C is 500 μm rostral to B). CTβ is transported avidly in both the retrograde and anterograde directions, however, and in this case there is robust retrograde labeling of neurons in the accessory basal nucleus of the amgdala (AB, arrow in inset in panel A) and scattered labeling throughout the basal-accessory basal complex of the amygdala. The AB and other parts of the basal complex project robustly to the accumbens shell (Krettek and Price, 1978; Brog et al., 1993). Thus, accessory basal neurons transported CTβ retrogradely from the injection site in the CeA and then anterogradely to the AcbSh. Additional abbreviations: ac—anterior commissure; VP—ventral pallidum. (Reprinted from Zahm, 2006, with permission.)

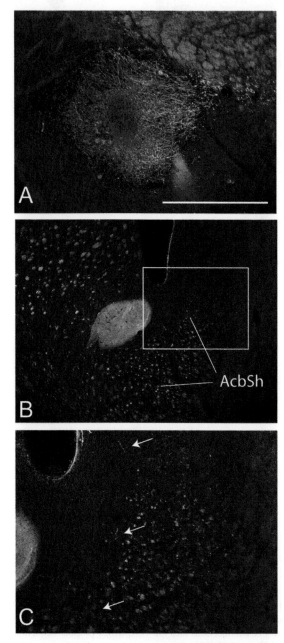

BSBOX 8 FIGURE B Micrographs showing sections illuminated in dark-field mode from a case in which a PHA-L injection was made in the central nucleus of the amygdala (CeA, shown in panel A). Panels B and C illustrate the accumbens shell (AcbSh) where negligible anterograde labeling (arrows) is observed. The box in B is enlarged in C. Scale bar: 1mm in A, 0.4mm in C. (Modified from Zahm, 2006, with permission.)

Printed and bound by CPI Group (UK) Ltd, Croydon, CR0 4YY

03/10/2024

01040316-0003